Praise for *The Poincaré Conjecture*

"O'Shea's approach is more narrative than technical, but he conveys the gist of topology's mind-bending contortions with great flair." —***New Scientist***

"The Poincaré conjecture is a guess about these higher-dimensional surfaces. If these ideas are hard to imagine—and they are—they are harder yet to prove . . . In *The Poincaré Conjecture,* Mr. O'Shea tells the fascinating story of this mathematical mystery and its solution by the eccentric Mr. Perelman."

—***Wall Street Journal***

"O'Shea, a mathematician, recounts the rich history of the famous problem and its solution, even managing to connect its abstract implications to the real world. A layman's guide to this mathematical odyssey is long overdue, and this one will appeal to math whizzes and interested novices alike."

—***Discover***

"At the genial center of this story is Poincaré, who made his mark by unifying geometrical, topological and algebraic approaches to vital questions in math and physics. O'Shea describes how Poincaré developed his famous question . . . For a lay person, it takes the rest of the book to do justice to the full statement of the conjecture, and it is worth the effort." —***Chicago Tribune***

"Donal O'Shea tells the whole story in this book, neatly interweaving his main theme with the history of ideas about our planet and universe. There is good coverage of all the main personalities involved, each one set in the social and academic context of his time."

—***New York Sun***

"Donal O'Shea, an accomplished topologist, tells the story of Perelman's proof, and what led up to it, for a general audience . . . *The Poincaré Conjecture* makes one of the most important developments in today's mathematics accessible to a wide audience, and it deserves to be widely read."

—***Toronto Globe and Mail***

"Donal O'Shea's beautifully written book, *The Poincaré Conjecture,* begins with a quick tour of the history of ideas of geometry and topology that set the stage for Poincaré. [O'Shea] has a gift for metaphor and often is able to capture the essence of a mathematical development or idea without falling into the jargon of the discipline or impenetrable prose . . . *The Poincaré Conjecture* is a fine example of mathematical writing for a general audience."

—***American Scientist***

"Even if I felt that after the third chapter I had the sort of mind powerful enough to bend spoons, I was still dazzled by what I was learning."

—**Bookslut.com**

"Getting from a simple question asked in 1904 to an elegant answer about a hundred years later is no easy task, but O'Shea (mathematics, Mount Holyoke College) makes the journey accessible to general readers and interesting to the mathematically inclined . . . Along the way O'Shea details the progress made in mathematics in the last hundred years of mathematics by a group of wildly dissimilar people."

—***Book News***

"O'Shea describes mind-bending structures in topology as clearly as most of us can describe a cube."

—***Publishers Weekly***

"O'Shea inspires readers to note the beauty, application, and humanity involved with this mathematical journey . . . [and] successfully weaves mathematical proofs with curious insights to tell a great story."

—***Library Journal***

"Donal O'Shea has written a truly marvelous book. Not only does he explain the long-unsolved, beautiful Poincaré conjecture, he also makes clear how the Russian mathematician Grigory Perelman finally solved it. Around this drama O'Shea weaves a tapestry of elementary topology and astonishing concepts, such as the Ricci flow, that have contributed to Perelman's brilliant achievement. One can't read *The Poincaré Conjecture* without an overwhelming awe at the infinite depths and richness of a mathematical realm not made by us."

—**Martin Gardner, author of *The Annotated Alice* and *Aha! Insight***

"The history of the Poincaré conjecture is the story of one of the most important areas of modern mathematics. Donal O'Shea tells that story in a delightful and informative way—the concepts, the issues, and the people who made everything happen. I recommend it highly."

—**Keith Devlin, author of *The Millennium Problems***

The Poincaré Conjecture

In Search of the Shape of the Universe

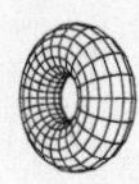
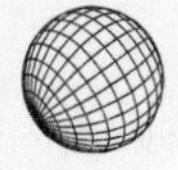

DONAL O'SHEA

WALKER & COMPANY
NEW YORK

Published by Walker Publishing Company, Inc., New York
Distributed to the trade by Holtzbrinck Publishers

All papers used by Walker & Company are natural, recyclable products made from wood grown in well-managed forests. The manufacturing processes conform to the environmental regulations of the country of origin.

LIBRARY OF CONGRESS CATALOGING-IN-PUBLICATION DATA HAS BEEN APPLIED FOR.

ISBN-10: 0-8027-1532-X (hardcover)
ISBN-13: 978-0-8027-1532-6 (hardcover)

Visit Walker & Company's Web site at www.walkerbooks.com

First published by Walker & Company in 2007
This paperback edition published in 2008

Paperback ISBN-10: 0-8027-1654-7
ISBN-13: 978-0-0827-1654-5

3 5 7 9 10 8 6 4

Designed by Rachel Reiss
Typeset by Westchester Book Group
Printed in the United States of America by Quebecor World Fairfield

Contents

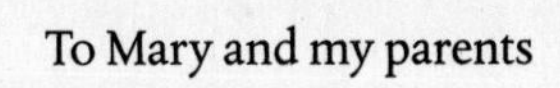

To Mary and my parents

Preface

This book is about a single problem. Formulated by a brilliant French mathematician, Henri Poincaré, over one hundred years ago, the problem has fascinated and vexed mathematicians ever since. It has, very recently, been solved. The Poincaré conjecture addresses objects that are central to our understanding of ourselves and the universe in which we live.

I write for the curious individual who remembers a little high school geometry, but not much more—although I hope that those with substantial mathematical backgrounds will also enjoy the book. Endnotes are addressed to those who know more, or who want to know more.

Go to any gathering. Sit beside anyone in a plane. And listen to what they say about mathematics. A few love it. But more do not, and what they say is not pretty. Some believe themselves congenitally incapable of mastering it. Some dislike it. And many loathe it with the passion reserved for love gone bad.

How can one subject, filled with so much beauty, inspire such a range of responses? The aversion some people feel seems to be rooted in fear. I have no illusions that one book will change this. But if you are a reader with ambivalent feelings toward mathematics, I hope this book inspires you to read further or, if you are a student or someone contemplating further study, to consider taking more mathematics courses.

I hope that you enjoy reading what follows as much as I enjoyed writing it.

1

Cambridge, April 2003

Revolutions in mathematics are quiet affairs. No clashing armies and no guns. Brief news stories far from the front page. Unprepossessing. Just like the raw damp Monday afternoon of April 7, 2003, in Cambridge, Massachusetts.

Young and old crowded the lecture theater at the Massachusetts Institute of Technology (MIT). They sat on the floor and in the aisles, and stood at the back. The speaker, Russian mathematician Grigory Perelman, wore a rumpled dark suit and sneakers, and paced while he was introduced. Bearded and balding, with thick eyebrows and intense dark eyes, he tested the microphone and started hesitantly: "I'm not good at talking linearly, so I intend to sacrifice clarity for liveliness." Amusement rippled through the audience, and the lecture began. He picked up a huge piece of white chalk, and wrote out a short, twenty-year-old mathematical equation.[1] The equation, called the *Ricci flow equation*, treats the curvature of space as if it were an exotic type of heat, akin to molten lava, flowing from more highly curved regions and seeking to spread itself out over regions with lesser curvature.

Perelman invited the audience to imagine our universe as an element in the gigantic abstract mathematical set of all possible universes. He reinterpreted the equation as describing these potential universes moving as if they were drops of water running down enormous hills within a giant landscape. As each element moves, the curvature varies within the universe it represents and it approaches fixed values in some regions. In most cases, the universes develop nice geometries, some the standard Euclidean geometry we studied in school, some very different. But certain tracks that lead downhill bring problems—the elements moving along them develop mathematically malignant regions that pinch off, or worse. No matter, the speaker asserted, we can divert such tracks; and he sketched how.

The audience had been drawn to the lecture by a paper Perelman posted to a Web site the previous November. In the paper's final section, he'd outlined an argument that, if valid, would prove one of the most famous, most elusive, and most beautiful conjectures in all of mathematics. Posed in 1904 by Henri Poincaré, the leading mathematician of his era and among the most gifted of all time, the Poincaré conjecture is a bold guess about nothing less than the potential shape of our own universe. But it is a guess, nonetheless. The challenge of proving or disproving it has exerted a sirenlike pull on mathematicians and has made it the most famous problem in not only geometry and topology but arguably in all of mathematics.[2] In May 2000, the Clay Institute, an institute dedicated to the advancement and dissemination of mathematical knowledge, had designated the problem as one of seven millennium problems and offered a million-dollar reward for its solution.[3]

Well over half the audience would have tried to make some headway on the Poincaré conjecture. Every person in the room—from the preppy-looking thirty-something-year-old with spiked hair, taking notes in Chinese, and the young blonde with the tight blouse and the too-short skirt, to the jogger in baggy running shorts and damp T-shirt, and the rheumy-eyed octogenarian with herringbone coat stained by decades of chalk dust—knew that they were potentially witnessing a monumental milestone in a three-thousand-year-old legacy. The mathematics invoked had been passed painstakingly from one era to another, through times of great wealth and times of shattering poverty, from the unknown Babylonian who gave us the area of a circle, to the perfect severities of Euclid, to the blossoming of geometry and topology over the last two centuries.

Two weeks and several lectures later, at the flagship campus in Stony Brook of the State University of New York a similar scene ensued. That lecture theater was even more crowded. This time, several reporters were in the room. The reporters had heard that Perelman had made a stunning discovery relating to the shape of our universe, and that he might win a million-dollar prize as a result. They had also heard of his shadowy career path—how he dropped out of sight a decade earlier, his brilliance recognized and his promise unrealized. A lightbulb flashed. "Don't do that," Perelman snapped, visibly annoyed.

The mathematician patiently answered all questions from the audience after his lecture. Those questions came furiously. "But that solution will blow up in finite time," came a voice from the middle of the room. "It doesn't matter," replied Perelman; "we can cut it out and restart the flow." Silence, then a couple of nods. The listeners were cautious, weighing what they heard. They would have to ponder his words for months to come, but this sounded promising.

Much of the mathematics Perelman called upon would have been inconceivable three decades previously. The technical tools that he used were at the very edge of what was possible and depended critically on the work of a number of persons in the audience. The atmosphere was tense. Everyone knew how delicate and how subtle the speaker's arguments were, and how easy it was to go astray. Everyone wanted them to hold. A Web site had sprung up,[4] managed by two professors, Bruce Kleiner and John Lott, in the University of Michigan's extraordinary mathematics department. The site contained links to Perelman's papers. Mathematicians around the world had added remarks and arguments clarifying obscure points and expanding on passages that seemed too terse.

Almost every mathematician, geometer or not, knew someone at the lecture and was waiting for an account. Most of the audience took notes, for their own benefit and for friends. Two, Christina Sormani, a young professor at Lehmann College, and Yair Minsky, then a freshly minted full-professor at Yale, would post their notes on the Web site, so that others would have access.

As at MIT, everyone in the room, young and old, except the reporters, realized that what they were hearing was the culmination of over a century of the greatest flowering of mathematical thought in our species' history. The lecture demanded close attention, leaving little space for stray thoughts. Even so, many would have thought of an especially cherished event or paper, recent or long past, that relates to Poincaré's work or of a person, possibly long dead, who would have loved to have heard this talk. All delighted in the wealth of good ideas and the promising paths to be explored.

The reporters, on the other hand, wanted to know about the million dollars. How did Perelman feel about the possibility of winning that kind of money? As it dawned on them that he did not care, they changed their approach and wrote stories about a reclusive Russian making a big math discovery, and speculated that he would reject the prize. Perelman filled in more details on subsequent days and during hastily organized discussion sessions.

But he refused all interviews with journalists and returned to Saint Petersburg a few weeks later without responding to the job offers proffered by top American universities.

The Poincaré conjecture and Perelman's proof of it is one of the greatest achievements of our age; it tells us much about the possible nature and shape of our universe. The Ricci-flow equation Perelman wrote, a type of heat equation, is a distant relative of the Black-Scholes equation that bond traders around the world use to price stock and bond options. But curvature is more complicated than temperature or money. As subsequent chapters will explain, curvature is a geometric object that requires more than one number to describe it, and the Ricci-flow equation that Perelman uses is shorthand for six linked equations, a triumph of elegance, simple to behold and concealing dazzling riches. Its closest analogue is the Einstein equation of general relativity that expresses the curvature of space-time.

The Poincaré Conjecture tells the story of the mathematics behind the conjecture and its proof. To speak of mathematics sensibly is to speak not just of results, but of the people who brought those results to pass. To the extent that mathematical achievement enters popular consciousness at all, it often reflects the romantic myth of a solitary genius heroically wresting understanding from an uncaring cosmos. There are indeed individuals whose insights seem to come from nowhere and who have single-handedly moved the discipline decades ahead. But, as colorful and as mysterious as genius is, mathematical progress also depends on thousands of other individuals, and on the institutions and societies in which they work and live. It is past time to tell this larger story. The narrative ranges from Babylon of five thousand years ago to present-day Saint Petersburg, northern New York State, and Madrid. It tells of exploration, war, scientific societies, and the emergence of the research university in Germany and most recently in the United States—tracing the history of geometry, the discovery of non-Euclidean geometry, and the birth of topology and differential geometry through five millennia, dozens of societies and human institutions, and hundreds of individuals.

Mathematical exposition has been interwoven with biographical, cultural, and historical material. There will be too much mathematics for some and far too little for others, but most people with a high school education will be able to follow the basic concepts of this book, even if the finer points are a bit challenging. One can understand and appreciate the mathematics and this famous conjecture without being able to "do the math" oneself. For the convenience of readers, a glossary of mathematical terms, an index of persons, and a list of major events detailed in the narrative follow the endnotes.

Some of the mathematics originated deep in the past, millennia ago. Mathematical investigation is one of the oldest human activities, as old as the trades of carpentry, cookery, and metal-working. But in fact, more mathematics has been discovered since 1900 than in all previous human history. As a result, the pace picks up and the reliance on endnotes for details and references necessarily increases as the narrative advances closer to the present. Lightly skim the more mathematical sections and notes. There is no test. You can always return later, if you wish, to puzzle out whatever seems unclear. After all, Poincaré's conjecture has stumped the most learned mathematicians for the last hundred years.

2

The Shape of the Earth

The Poincaré conjecture provides conceptual and mathematical tools to think about the possible shape of the universe. But let us start with the simpler question of the shape of our Earth. Any schoolchild will say that the Earth is round, shaped like a sphere. And, in these days of airplanes and orbiting spacecraft that can take pictures of our planet from on high, this seems utterly obvious. But, in times past, it was difficult to say with certainty what the shape of the world was.

"Is there anyone so foolish as to believe that there are persons on the other side of the earth with their feet opposite to ours: people who walk with their heels upward and their heads hanging down?" According to Washington Irving, mid-nineteenth-century America's celebrated public intellectual, this rhetorical question of an early Christian Church father was among those cited by the advisory committee convened by King Ferdinand and Queen Isabella to evaluate Christopher Columbus's proposal to sail west to the East Indies.[5] Irving breathlessly recounts how the skeptical, even hostile, committee members, blinded by their belief in a flat Earth, repeatedly challenged Columbus.[6] Alone before the learned committee, in the land of the Inquisition, Columbus steadfastly defended his views.

Irving's portrait of Columbus has endured and has been retold uncritically by generations, but it is nonsense—"pure moonshine," wrote the highly respected American historian Samuel Eliot Morison.[7]

In 1490, virtually every educated Westerner believed that the Earth was spherical. There were, to be sure, debates about the existence of antipodeans, persons who lived on the other side of the world and who could not have known Christ or the Prophet. In the absence of knowledge, fantasy flourished. There were widely believed tales of huge areas with fearsome storms that made passage impossible. Horrific depictions of monsters abounded. Some argued that the other side was a great sea with no land and no antipodeans.

Columbus and the advisors to Ferdinand and Isabella did disagree, but their disagreement was about the size of the Earth, not its shape. All of them believed that the world was spherical. However, although there were some very good, detailed maps of parts of the Earth (notably the Mediterranean basin), and assemblages of these maps into atlases, no one really had any idea of the distance around the world. The most authoritative estimates had come from the ancient Greeks. In the second century, Ptolemy estimated the circumference as 18,000 miles. The court advisors of the Spanish monarchs favored the estimate of the Erastosthenes, a Greek geometer of the third century BCE: He had given the circumference a value of 24,200 miles, very close to today's value of 24,902 miles. Columbus argued that the Earth was even smaller than Ptolemy thought. The advisors, not Columbus, were right. Had their views prevailed, Columbus would not have received the financial support he needed, because the cost of outfitting an expedition for the longer voyage and the increased risk would have been prohibitive.

Columbus's reputation has risen and fallen over time. He was hailed initially for his wisdom, his courage, his vision, even his looks. The five-hundredth anniversary of his voyage brought some different, darker assessments: Columbus as rapacious imperialist, headstrong, the beneficiary of blind luck. The explorer was indeed lucky that the Americas were where they were. However, he was acting on the best information then available. He had heard of rumors of the Norse voyages and those of the Irish navigator, Brendan. If they had reached Asia, as it then seemed reasonable to suppose, then Erastosthenes and Ptolemy both had to be wrong. This was certainly more reasonable than supposing there was an undiscovered continent between Europe and Asia: there were abundant instances of data upon which Ptolemy relied having been erroneous.

To his death, Columbus believed that he had reached the Spice Islands to the east of India. He knew that it took far longer to get to them by sailing around Africa and he struggled to reconcile this with his own observations. He writes: "I find that the earth is not as round as it is described, but it is shaped like a pear, which is round everywhere except near the stalk where it projects strongly; or it is like a very round ball with something like a woman's nipple in one place, and this projecting part is highest and nearest heaven."

This passage has often been ridiculed, but I find it quite inspiring.[8] Here we have an old man, who has believed and argued all his life that the Earth is a perfectly round sphere, still open to other hypotheses that better fit the data. Perhaps, he is arguing, the Northern Hemisphere is like a small narrow stalk and the Southern Hemisphere bulges like the bottom of the pear. Maybe we can reach the Spice Islands relatively quickly by sailing around the narrow neck of the Northern Hemisphere, whereas to sail around Africa in the larger Southern Hemisphere to get to them results in having to travel much further. The willingness to reexamine lifelong beliefs because of conflicting data takes enormous courage, and contrasts sharply with recent examples of public discourse in which our political, cultural, and religious leaders have fit data to preconceived theories.

Arguments about the circumference of the Earth aside, the fact is that no one in 1490 really knew whether the Earth was finite, much less a sphere. About the only thing known for certain was that the Earth was curved, and large regions had been mapped. Where, then, did the belief that the world was a sphere arise, and how did it gain such wide acceptance?

IONIA AND THE GREEKS

The story of how the Earth was first understood to be spherical begins about two thousand years before Columbus's voyage, on the Greek island of Samos. In Columbus's time, Samos was almost entirely depopulated. Its location a mile off Turkey's western coast had made it a target of every possible invader: Byzantines, Arabs, Venetians, Crusaders, Turks. Even today, the quiet towns, the white sand beaches, the olive trees and the vineyards along the road circling the Karvounis or Kerkis mountains, suggest nothing so much as warm indolence and ageless torpor. But drive or cycle eight miles south of the present-day town of Samos and one will come to the town of Pythagoreion. Named for Pythagoras, Samos's most famous citizen, it sits on the partially buried ancient city of Samos. Take the left fork in the road and continue along the seashore. Over a rise is a valley and the ruins of the Heraion, the temple of Hera, one of the seven wonders of the ancient world. A single column, half its original height, stands on a massive stone foundation, all that remains of the 155 columns that once graced this place. Nearby is the Evpalinos aqueduct-tunnel, now strung with lights and open to the public. The sense of grandeur past overpowers the pallid present.

Continuing another 20 miles from the ruins of the temple along the main road brings the hilltop town of Marathokampos into view to the west. A small sign there marks a hiking trail up Mount Kerkis and to a cave where Pythagoras taught. A moderately strenuous ascent leads to a shaded grotto and a spacious cave. Walk down a bit, and the rocky slope of Kerkis and the impossibly blue Aegean Sea stretch out far below. The remoteness, the beauty, and the solitude call to mind Ireland's Skellig Rock or the eastern slope of the sacred volcano Haleakala on the Hawaiian island of Maui. At such places, the present thins and the past seems close. Ancestral voices murmur just below the threshold of hearing. It is here that Pythagoras first taught that the Earth was a sphere. His spirit is still here, the locals say. On dark nights when the winds scour the face of Kerkis, they tell of the faint light that shines out from the rocks guiding mariners far below.

That light burned brightest twenty-five hundred years ago in the heyday of Samos, then an important city-state of Ionia. This small region consisted of the westernmost coast of Turkey, or Asia Minor, extending from Phocaea south about a hundred miles to Miletus, together with the islands in the Aegean just off the mainland. Oral tradition maintains that the Greeks had colonized the region at the beginning of the first millennium BCE. Here, the Greek economy and culture rebounded after the so-called Dark Age, the period of nearly five hundred years of depopulation, economic hardship, and the loss of literacy following the violent and still mysterious destruction of the Mycenaen civilization in the twelfth century BCE.

The work of the Ionians Homer and Hesiod marked the beginning of a renaissance. Writing, based on the alphabet from Phoenicia (present-day Lebanon) was introduced in the late ninth century BCE and promptly used to write down Homer's epics. Between the seventh and fifth centuries BCE, Samos became a great naval power, and the mainland Ionian city-states important and powerful trading centers. Greek philosophy and science emerged in Ionia at the same time.

Ionia's location on the fringes of Asia Minor meant that Ionian thinkers had contact with the other great civilizations of the eastern Mediterranean, particularly Egypt. Far to the east was the legendary Babylon and, further still,

FIGURE 1. *Ionia*

Persia. The great Ionian philosophers Thales (624–547 BCE) and Anaximander (611–545 BCE), both of Miletus, taught that the cosmos and the motions of the stars were governed by laws of nature, not magic or the arbitrary intervention of divine beings. The cosmos was ordered and could be understood through reason and logical thought. This was a new idea, one that would take a while to catch hold, and one whose influence would wax and wane in different times and places.

Thales and Anaximander also speculated on the shape of the Earth. Thales had adopted the Egyptian belief that it rose like a hill from a smooth, presumably infinite, sea. Anaximander, who had given the matter a little more thought, believed that the Earth was shaped like a cylinder and suspended in space.

Pythagoras (569–475 BCE) had a much larger appetite for religion and mystery than did the Ionian philosphers. His father, Mnesarchus, a trader from the wealthy city-state of Tyre, moved from Phoenicia to Ionia, where he met and married Pythais of Samos. According to one story, Mnesarchus was granted citizenship on Samos after having brought grain to the island during a time of famine. Pythagoras traveled widely with his father, meeting Syrian and Chaldean scholars on a trip back to Tyre, and journeying to Italy and

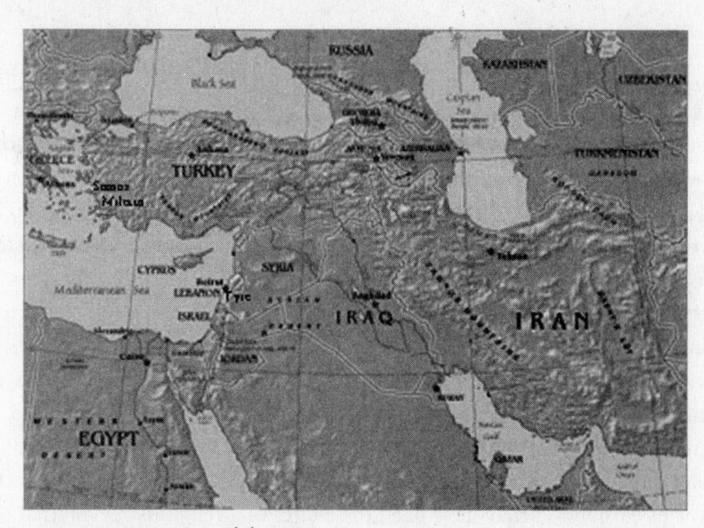

FIGURE 2. *Ionia, Egypt, and the East*

Greece. He was something of a prodigy and displayed an early interest in philosophy that was nurtured by his teacher, Pherekydes (c. 600–550 BCE), and by Thales and Anaximander. Thales made an especially strong impression; famous and respected throughout Ionia, he was an old man when he met Pythagoras. Thales had spent time in Egypt in his youth and it was he, fatefully, who encouraged Pythagoras to do the same.[9]

In Egypt, Pythagoras somehow managed to become initiated into the sacred Egyptian mysteries. Just how a foreigner was allowed to learn the highly secret rites is unclear. All sources agree, however, that he had a striking golden birthmark on his leg, which suggests, perhaps, that the Egyptian priests believed him to be favored by their god Osiris and thus allowed him to join their priesthood. Certainly, there were widespread rumors later in Pythagoras's life that he was partly divine and had been touched by Osiris, rumors that he appears to have done little to discourage. Details of his years in Egypt are even more muddled than those of the rest of his life. He seems to have been captured, presumably in the Persian invasion of Egypt in 525 BCE, and taken as a prisoner from Egypt to far off Babylon, 55 miles from present-day Baghdad and then the richest city in the world. There he learned the mysteries of Persian dualism, absorbing the teachings of Zarathustra (known to the Greeks as

Zoroaster). Pythagoras must also have learned much of his mathematics in Babylon. Babylonian mathematics was much further advanced than Egyptian mathematics (which had come from Babylonia) and, although there is a fair amount of debate about how much mathematics Thales knew, Pythagoras knew much more and attained a much higher level.[10]

The circumstances of Pythagoras's release from Babylon are lost, but his return to Samos did not go unreported. A striking figure in his Eastern trousers and garb, worn to conceal the birthmark, and a spell-binding orator, he was a sensation. His teachings thoroughly interwove the rational and irrational, the scientific and the mystical, and provided a marked contrast to the sobriety of the Ionian philosophical school. Their power still inhabits the abandoned surrounds of his cave on Samos.

Around 530 BCE, Pythagoras and a number of followers moved to Croton, a quiet Greek colony in the south of Italy that had been founded nearly two centuries earlier and may have been the center of a religious revival that was sweeping the region. There, Pythagoras founded his school. Actually, it was less a school than a brotherhood, itself a misnomer as this association of like-minded truth-seekers also admitted women. Called "the semicircle," it had an inner group of men and women, the *mathematikoi*, strict vegetarians who lived communally, did without personal possessions, and were taught by the great man himself; and an outer group, the *akousmatics*, who lived in their own houses under much less stringent rules. Members underwent elaborate initiation rites and were sworn to secrecy.

The Pythagoreans believed that, at its deepest level, reality is mathematical, that all beings are related, that philosophy can be used as a means for spiritual purification, and that the soul can rise to union with the divine. The attractiveness of their notion of universal relatedness and the exotic blend of Eastern mysticism and Greek thought fascinated their contemporaries. Pythagoras and the Pythagoreans flourished, and became known throughout the Greek world. Time has softened the debates that raged about Pythagoras's character and stilled the legends that swirled about the man. He was a riveting speaker, alleged to have remembered details from previous incarnations. To admirers, he was a genius with extraordinarily deep knowledge, wise and compassionate beyond measure. To detractors, he was a charlatan with an unerring instinct for self-promotion. Whatever the truth, Pythagoras was enormously influential, and the prestige of his brotherhood actually increased after his death.

Most important, Pythagoras had taught that the Earth was spherical. He had begun to accumulate evidence to support this idea, and had been the first to conceive of the Earth as existing together with the stars in a single universe. Later Pythagoreans, notably Philolaus (c. 470–385 BCE), even relinquished the notion of a geocentric universe, teaching instead that the Earth, Sun, and stars orbited an unseen central fire.

FROM PYTHAGORAS TO COLUMBUS

So now we know when and where the notion that the surface of the Earth was a sphere originated. If the Pythagoreans were sworn to secrecy, how did their knowledge leak out? And once it did, why was it taken seriously and how did it get transmitted to modern times?

The first thing to realize is human nature has changed little over three millennia. Mystery attracts us, and few resist the allure of great knowledge secretly held. Witness the success of recent novel and film *The Da Vinci Code*. The visibility of the Pythagoreans guaranteed a market for books purporting to expose their concepts and beliefs. And lots appeared. Philolaus allegedly wrote his book *On Nature* because he needed money.

The great philosopher Plato (427–347 BCE) would be much influenced by the Pythagoreans. He purchased Philolaus's book for the Academy, the school he founded in Athens. He also had been a very close friend of the first-rate Pythagorean mathematician, Archytas (c. 428–350 BCE). As a result, many Pythagorean ideas made their way into Plato's system and the mainstream of Greek thought. The Academy's most brilliant student, Aristotle (384–322 BCE), though less enamored of the Pythagoreans (he is reputed to have dismissed them as "filthy vegetarians"), incorporated many of their ideas into his own formidable body of work.

The importance of Aristotle is difficult to overstate, for his intellectual reach was enormous. He codified the rules of formal logic, systematized philosophy, and contributed to all of the natural sciences. He taught that the Earth was a sphere around which the Sun and Moon rotated. His works on ethics, aesthetics, and politics are still read today. Medieval thought, both Christian and Islamic, is rooted in Aristotelian principles. Backed by the authority of Plato and Aristotle, and by discoveries that I shall describe momentarily, the notion that the earth was spherical took firm hold.

Aristotle's role as private tutor of Alexander the Great,[11] then the most powerful man in the world, served to propagate Pythagorean ideas even more decisively, albeit less straightforwardly, than his direct teachings. Alexander had built on his father's conquest of Greece in 338 BCE and had gone on to conquer the then-known world. By the time of his death in 323, Alexander's empire extended from around the Mediterranean into India.

Although the complexities of the relationship between the tutor and his gifted student are forever lost, Aristotle doubtless importuned Alexander to use his campaigns for scientific observation. Whether on account of this intercession or not, Alexander took mapmakers on his campaigns. Their maps did not survive, but the writing of some of his generals did, and their descriptions and accounts became the basis of maps created centuries later.

After Alexander's death, his empire fell apart. The largest share fell to his general, Ptolemy Soter I, who chose as his capital Alexandria at the mouth of the Nile. Here, in the first, and the greatest, of the eponymous cities commissioned by Alexander the Great, Ptolemy began the construction of the fabled library that would secure Alexandria's position as the intellectual and cultural capital of the world. Scholars flocked to the great library, eager to do research in a collection that held hundreds of thousands of books and scrolls. The position of head librarian was perhaps the highest academic position that one could hold in the ancient world, easily rivaling the headship of Plato's Academy in Athens and akin to the presidency of Harvard University or mastership of Trinity College today.

Eratosthenes (275–195 BCE) of Cyrene (now Shahhat in Libya) became the third head librarian in 235 BCE in the reign of Ptolemy II. A geographer who had studied in Athens, Eratosthenes also wrote poetry and literary criticism, and pursued work in mathematics, astronomy, and philosophy. He had no doubt that our world is a sphere—if one were to sail west from Spain, he wrote, one would eventually encounter India. He even made a map of it.

Most famously, he noted that two individuals many miles apart, one directly north of the other, will see the Sun at different angle at the same time of the day, and pointed out that if one assumed that the world was a sphere, one could use the differences in the angles to estimate the distance around the world. Eratosthenes measured the difference between the angles between the Sun at noon at Alexandria and at what is now Aswan, 486 miles south on the Nile, and produced the stunningly accurate estimate[12] of the circumference of the Earth championed by some of Columbus's detractors at the Spanish court. A few generations later, Hipparchus (190–120 BCE), proposed that the Earth circled the Sun, and established measurement by latitude and longitude, dividing the world into 360 degrees.

Alexandria continued to be home to geographers, mathematicians, and astronomers long after the great library was destroyed.[13] The greatest of these was Claudius Ptolemy (85–165 CE). His *Geography* was a category killer: a book that collated all previous knowledge, that was utterly authoritative, and that became the absolute standard. Ptolemy discussed the problem of mapping a curved earth onto a flat piece of paper and of dealing with the possible projections of a sphere to a plane. He pointed out that once one has associated coordinates (in this case latitude and longitude) with places being mapped, one can reconstruct a map at will. Using the surviving writings of Alexander's generals and data acquired by other travelers, Ptolemy calculated the latitude and longitude of all known places.

The maps included in the original edition of Ptolemy's *Geography* have been lost. However, that barely matters. The text is absolutely lucid and can be read with profit even today.[14] Overall, Ptolemy imagined that the habitable portion of the Earth, from the coast of Western Europe to India and beyond, was about half of the entire planet. He estimated the circumference of Earth at 18,000 miles, considerably less than the value given by Eratosthenes, but greater than that of Columbus.

Ptolemy's *Geography* was largely forgotten for many years, except by a number of Muslim scientists. In Palermo, in the multicultural court of the Norman King Roger II, al-Idrisi (c. 1100–1165) used an Arabic translation of the great work and improved on Ptolemy's calculations. The Greek text was lost and not rediscovered until a Byzantine monk, Maximos Planudes (c. 1260–1330), found a manuscript copy without the maps. Planudes reconstructed some maps and commissioned the re-creation of others. In 1406, the text was translated into Latin, and a Benedictine monk, Nicolas Germanus, redrew the maps on a trapezoidal projection, one of the three proposed by Ptolemy.[15] This became the basis of the first printed Ptolemy atlas, which was published in Bologna in 1477 in an edition with five hundred copies. Columbus owned a copy and studied it carefully.

The transmission of Pythagoras's view, via Plato, Aristotle, the learned geographers of Alexandria, Sicily, and the early medieval world, had been successful. By the time of Columbus, virtually everyone believed that the Earth was spherical. The preponderance of the evidence supported this belief. If one looked at the Sun from different places on a north–south line, one saw it at different angles. When a ship sailed into sight from a long distance off, one could first make out its mast, then the rest of the ship. The tides, night and day, the phases of the Moon, and many other natural phenomena could be more reasonably understood by thinking of the Earth as a sphere.

THE SHAPE OF THE WORLD

Belief is one thing, but when did we really know, without any doubt, that the world was shaped like a sphere? We have seen that Columbus began to doubt the sphere theory, thinking of the Earth as pear shaped. And today we know that our planet is not a perfectly round sphere, but is flattened somewhat at the poles. But, as we shall see presently, there are other more radical possibilities: the question of the Earth's shape is much more than a matter of bumpiness and flattened regions.

Our certain knowledge of the world's shape had to await exploration and careful mapping of all regions. The two centuries following Columbus saw the production of ever more extensive atlases of the Earth. These have been among the most prized and sought-after books of all time. The Ptolemy atlas that Columbus had used was republished in Rome in 1508 by Berandus Ventus de Vitalibus. This was the first edition to include the European voyages to the new world, and around the Cape of Good Hope. The world map included a small America.

Jacobus Pentus de Leucho, of Venice, republished Ptolemy as *Liber Geographicae* in 1511, with twenty-eight maps and the text carefully edited by Bernardus Sykvanus, of Eboli. Abraham Ortelius, geographer to King Philip II of Spain, published his *Theatrum Orbis Terrarum* in 1570. It was to go through many editions, but there were large blank spaces where nothing was known, and the scales were wildly inaccurate.

Ortelius's friend, Gerhard Mercator, of Rupelmonde, a Belgian and the greatest geographer since Ptolemy, revolutionized mapmaking by introducing a map projection (now known as the Mercator projection) that allowed mariners to plot constant bearing courses. Atlases based on his work and the continuing expeditions of European explorers began to appear at the end of the sixteenth century in Amsterdam. In fact, the word *atlas* referring to a collection of maps comes from Mercator's practice of prefacing his map collections with a picture of the Greek god Atlas, shown supporting the world on his shoulders.

The most expensive, and the most celebrated, printed book of the seventeenth century was Joan Blaue's multivolume *Atlas Major*, which was published in four languages in 1662–63. Stunningly beautiful, it nonetheless contained many inaccuracies (some unforgivable, even at that time) and gaps.

Although it seemed highly unlikely that the world went on forever or, worse, came to an end, we did not know for certain that neither hypothesis was true until the return in 1522 of Magellan's expedition around the world.[16] And even after Magellan, it was not completely clear that the Earth was a sphere. Other possibilities existed.

This may seem ridiculous. But is it? The extraordinary world map created by Battista Agnese in 1546 depicts Magellan's voyages. Agnese clearly thought the world was spherical. And in looking at this map today, we tacitly think of all points on the top edge as going to a single point (the North Pole), and all on the bottom edge as being identified with a single, different point (the South Pole). Moreover, each point on the right edge corresponds to a point on the left edge with the same latitude. Modern observers know that this is the "right" way to interpret this map, and others like it, because people have explored the world and know which areas are continuous lands or bodies of water. But in Agnese's time, with such large territories unmapped, it was conceivable that one might go off the north (that is, top) edge of the map and come back in the south (that is, bottom) edge of it. Or that maybe one could go north (or south) forever and never come back.

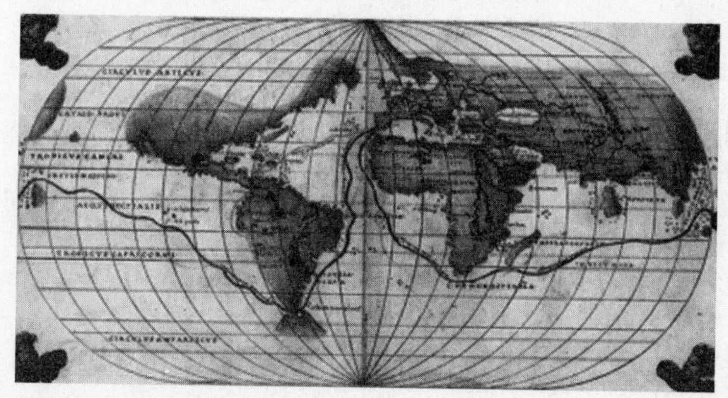

FIGURE 3. *Agnese's map*

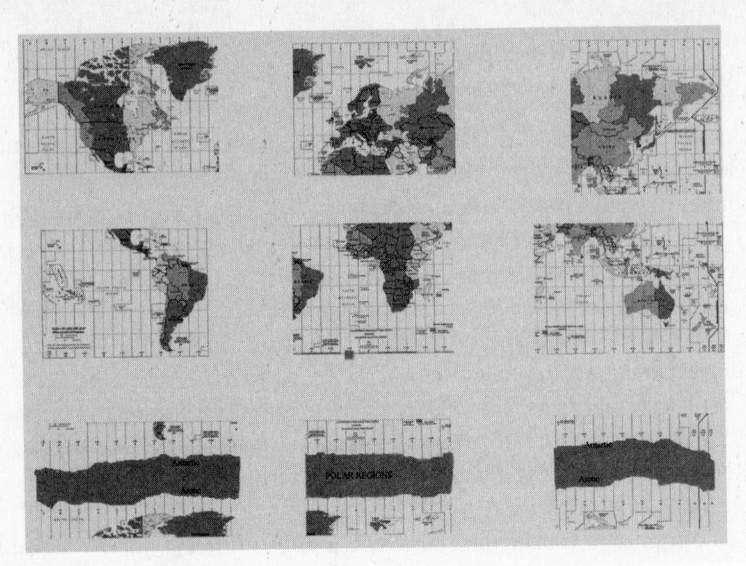

FIGURE 4. *Atlas of an imaginary world*

Consider, for example, figure 4, the atlas of an imaginary world that does not appear at first sight too different from ours. The maps do not fit together on a sphere!

To understand the world covered by these maps, connect the maps together by overlapping the eastern (rightmost) part of each map to the western (leftmost) part of the map on the right. Likewise, overlap the north (top) part of each map with south (bottom) part of the map immediately above it. So far so good. We thereby obtain the map of the world shown in figure 5.

Now, the places depicted on the right edge of the map are the same as those on the left edge. Again, so far so good. This is just like what we are used to. But we are also accustomed to assuming that all points on the top part of the map are a single point, the North Pole. Look closely. This is *not* the case here. Indeed, every point along the top edge is the same as a point at that same longitude along the bottom edge, but the points at different longitudes on the top edge are different.

To get a representation of this world in space, a globe so to speak, we have to glue the right edge to the left edge and the top edge to the bottom edge. What do we get? Now, as a schematic, figure 6, shows, when we glue the top

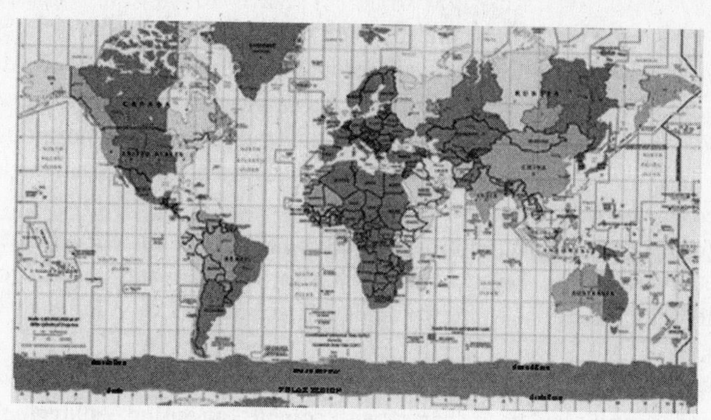

FIGURE 5. Gluing the maps to make a world map

edge of a rectangle to the bottom we get a cylinder. Gluing together the right and left edges amounts to attaching the right and left circles of the cylinder, which gives the surface of a doughnut. Such a surface is called a *torus* (to distinguish it from the solid thing that we eat).

One might object that there is no way that our world could look like a torus. If one lived on the surface of the inner ring (facing the hole in the doughnut), wouldn't one see the part of the surface opposite rapidly rising off into space? Maybe. But what if the world were enormous? Or what if the part

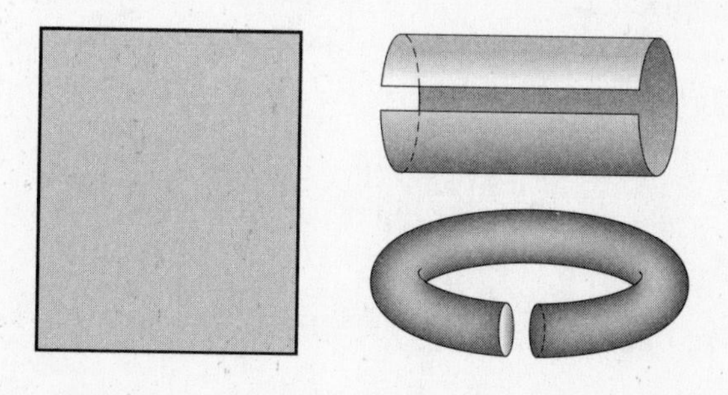

FIGURE 6. Connecting the top and bottom edges of the rectangle on the left gives the cylinder on the upper right. Connecting left and right edges then gives a torus. Try to imagine overlaying the map in figure 5 onto the torus.

along the inner ring corresponded to the polar regions in the world map in figure 5? How could we have ruled this out in Columbus's era? We could not have. Magellan could very well have circumnavigated the inner ring of a torus—or, indeed, the outer ring. And if the world were truly enormous, he may have just gone from outer to inner and back again, tracing what might be called a bite-size trajectory on the torus.

To return to the map in figure 3, suppose that it had turned out that our world was such that one could keep going north forever, and similarly to the south. In this case, our world would have been shaped like an infinitely long cylinder.

We conclude that we could not know the shape of our world with absolute certainty until it had been charted with complete precision—all regions including the poles. And the poles and interiors of some continents were not mapped until the nineteenth century.

3

Possible Worlds

Popular accounts of mathematics often stress the discipline's obsession with certainty, with proof. And mathematicians often tell jokes poking fun at their own insistence on precision. However, the quest for precision is far more than an end in itself. Precision allows one to reason sensibly about objects outside of ordinary experience. It is a tool for exploring possibility: about what might be, as well as what is.

The discussion in the last chapter raised the possibility that the world could have been a torus. We don't see doughnut shapes when we look up in the sky, so it takes some openness to possibilities to be willing to assume the Earth is one. Nowadays, we can get off the Earth and take a photograph from a satellite or spaceship. But in the era before space flight, it took an act of interpretative imagination to see the Moon and the Sun as spheres instead of flat disks facing us. The other planets and stars appear to be point sources. Rather than argue about what we are seeing in the heavens, let's suppose that we cannot see beyond the Earth. Imagine that we live on a planet like Venus that is covered in cloud all the time. What shape might our planet be? Could it be something else other than a sphere or a torus?

Although these questions have been framed by asking you to imagine worlds whose shapes we know to be "wrong" in the ordinary sense, it is characteristic of mathematics that, time and again, such feats of imagination lead to new understandings and new structures that later prove to be exactly what is needed for a major scientific advance.

To go further, we need terminology that is clear and unambiguous. The most critical idea for us will be that of a two-dimensional manifold or surface. We arrive at this notion by thinking about possible shapes that a world might have, and it is relatively harmless to imagine two-dimensional manifolds as modeling worlds on which we might live. In particular, let us agree that a *two-dimensional manifold* or *surface* is a mathematical object all areas of which can be represented on some map on a sheet of a paper. The word *two-dimensional* refers to the fact that at any point on such an object, nearby points can be expressed in terms of two independent directions. This is important because mapmaking requires that we can determine how points are related to one another. We must be able to identify each. The maps, that is, the sheets on paper on which the points of the world are represented, are two-dimensional. A collection of maps that covers the surface, so that every point on the surface is represented on at least one of the maps, is called an *atlas*. If one purchases an atlas of the earth, one will get a book of maps, and one rightly expects that every location on earth appears in at least one of the maps. A two-dimensional manifold or surface, then, is an object represented by an atlas.

Several remarks are in order. First, two-dimensional manifolds are mathematical objects that are idealizations of physical reality. When we say that the earth *is* a sphere, we are saying that the mathematical object that is a sphere is a good model for the surface of the earth. Notice, by the way, that by a *sphere,* we mean the outside skin, or surface, of a ball. We do not include the stuff enclosed by the sphere. So that when we say the earth is a sphere, we are excluding the bedrock or magma beneath the surface. Likewise, a *torus* is any two-dimensional manifold that models the surface or skin of a doughnut. It does not include the inside. The extra precision we get by making careful definitions allows some strange objects, and we must be careful not to confuse physical considerations and mathematical ones. For us, a two-dimensional manifold is a set with the property that all points near a point can be represented on a map. That's all. Mathematicians use the word *surface* as synonymous with *two-dimensional manifold,* although not every two-dimensional manifold is the surface of some solid.[17] Nor is it always possible to consistently define right and left on a manifold.[18] However, it turns out that any two-dimensional manifold on which it is possible to consistently define left and right can be represented as the surface of some solid, and vice versa. Such two-dimensional manifolds are called *orientable.*[19]

Another crucial thing to keep in mind is the use of the word *dimension*. In the casual usage of everyday speech, one will often hear assertions such as the Earth (or a sphere) or a torus is three-dimensional because one needs a three-dimensional space in which to fully fit it. We will *never* use *dimension* in this way. For us, *dimension* refers to the number of independent directions needed to represent all points on an object near a given point. If we tried to put all the maps in an atlas together to get a sort of globe that represented the surface more holistically, we will certainly need a third dimension (or more), but we still refer to the manifold or surface as being two-dimensional. The dimension refers to the number of independent directions that someone who lived on the manifold would experience, not the number of dimensions that we need to fit the object into.[20] So, the surface of the Earth is two-dimensional because, to represent a region of it, we use a map on a piece of paper (or, equivalently, we can use two numbers, such as latitude and longitude, to represent any point near a fixed point). A plane is two-dimensional, but any line, whether or not it is curved (and, in particular, a circle), is one-dimensional. The space in which our world exists (that is, our universe) is three-dimensional, as is the region beneath the surface (that is, the bedrock and magma) of the Earth. Later, we shall return to the notion of dimension, defining it even more carefully and reducing it to numbers. It will turn out that there will be manifolds of any dimension.

For now, we mention two other terms that are used in common language, but not precisely enough for our purposes. The first is *boundary*. Some two-dimensional manifolds have a boundary, and others do not. The boundary of a two-dimensional manifold is its edge, or collection of edges, as seen from the point of view of someone in the manifold. A plane that goes off infinitely in every direction does not have a boundary, but a disk in the plane has a boundary, namely the circle that is its boundary. The outside surface of a one-foot-long segment of a copper tube has a boundary—namely the circles, one at each end. A sphere has no boundary (although it is the boundary of the solid ball within). If you live on the Earth, you don't come to an edge where the earth ends. Similarly a torus has no boundary (although it is the boundary of a solid doughnut within it). If a two-dimensional manifold has a boundary, then that boundary is one-dimensional. The notion of a boundary goes over to objects with different dimensions. A circle has no boundary (even though it is the boundary of the disk within it). Nor does a straight line that goes off infinitely in both directions. But a straight line segment that is one foot long has a boundary, namely the two points at each end. The solid interior of the Earth has a boundary, namely the two-dimensional manifold that is a sphere. If a manifold has a boundary, its boundary will be one dimension lower.

The second term is *finite*. We say that a two-dimensional manifold is finite (or *compact*) if only a finite number of maps are needed to cover it. The Euclidean plane (henceforth called *Euclidean two-space* or simply *two-space*) that we learned about in high school that stretches out forever in two independent directions is a two-dimensional manifold that is not finite. Both a sphere and a torus are finite manifolds—one can't go off forever in one of them without eventually coming back near to where one started.

One common misconception is that for an object to be finite, it must have a boundary. The very first arguments about whether the Earth was finite often conceptualized the alternatives as being either the Earth's going on forever or its having an edge that one could fall off. It did not at first occur to people that the planet might be a sphere (or torus) and thus be both finite and have no boundary. Of course, for the Earth to be a sphere requires our accepting the seemingly nonsensical idea that people on the other side are walking with their feet pointing toward us. While no one has trouble accepting this these days, there are many who make the same error when talking about the universe. They assume that if the universe is finite, it must have a boundary (which would now be a two-dimensional object) that we can't go beyond. This is not so.

GEOMETRY AND TOPOLOGY

We have one final bit of housekeeping. We need to be precise about what it means to say that two manifolds are the *same*. As with everything else, this depends on one's point of view. Two objects can be same, or equivalent, in one sense, but different in another. In talking about shape, one is usually not concerned with features like size or distance that pertain to *geometry*, but about properties that are preserved under stretching and small deformations. Such properties belong to the domain of *topology*. Let us say that two surfaces are the *same topologically* if the points of one can be put into one-to-one correspondence with the points of the other so that nearby points correspond to nearby points (such correspondences are called *continuous*).[21] Two manifolds that are the same topologically are also said to be *homeomorphic,* and the one-to-one correspondence that establishes that they are the same is called a *homeomorphism.* Topology studies properties of surfaces (and other objects) that allow one to tell whether or not two surfaces (or other objects) are homeomorphic. Such properties are called *topological properties.*

Topological properties can be very different from geometric properties such as length and angle. Any two surfaces that can be deformed into one another by pulling and stretching (no tearing—because that can destroy continuity) are homeomorphic. Two spheres of different radii are homeomorphic. The surfaces in the figure below are all homeomorphic, and a topologist would view them all as spheres. In particular, a topologist might wonder what Columbus was worrying about! The pear-shaped world he envisaged is still a sphere as is the surface of an apple. To avoid confusion, we sometimes refer to the symmetric sphere pictured on the left in figure 7 as a *round sphere.* Columbus, in his pear reverie, was speculating that the earth is not a round sphere, but a sphere nonetheless.

Clearly, homeomorphism is a very crude notion of sameness. The "knotted" torus in figure 8 is homeomorphic to a regular torus, although there is no way to deform one continuously into the other (at least if you insist on not passing points of the torus through itself).[22] In fact, if we attach the top and bottom edges of the map in figure 5 to each other to get a cylinder and then knot up that cylinder before joining up its right and left edges, we get a knotted torus. The inhabitants of a cloud-covered world in which no shapes could be seen in the distance would not be able to determine, just from the atlas, if their world were knotted. If we lived on a world that was a knotted torus, we could cut the surface up into maps and reassemble them to get an ordinary torus.

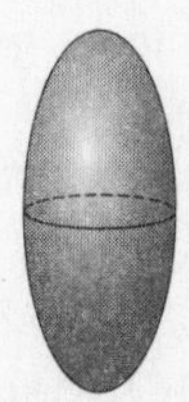
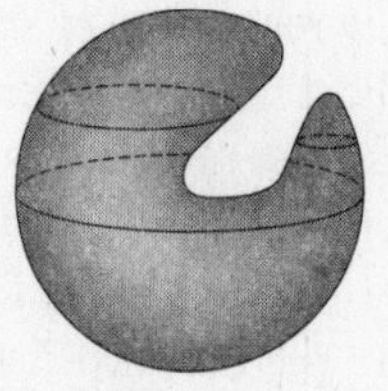

FIGURE 7. *These are all homeomorphic to the two-dimensional sphere.*

FIGURE 8. *Knotted torus*

CLASSIFICATION OF SURFACES

With the language we now have, we can ask some meaningful questions about what might be. Suppose that we live on a world in which we have mapped every region carefully. That is, suppose that we have a book full of maps that compose the whole world. Since there are a finite number of maps, we know that the world does not go on forever. If we redrew the maps to roughly the same scale, trimmed off overlaps, and tried fitting them together, what different types of shapes could we get? In other words, what are all possibilities for two-dimensional manifolds?

Let us first note that a sphere and a torus are not the only possibilities. Consider the *two-holed torus* sketched in figure 9. If we lived on such a world, and if we started off a journey from any point on it and kept going, we would never run off it. There are no edges. We could make a map of any region of it. And we could collect all the maps together in a book and create an atlas of the world. Would a geographer on this world examining the atlas be able to determine

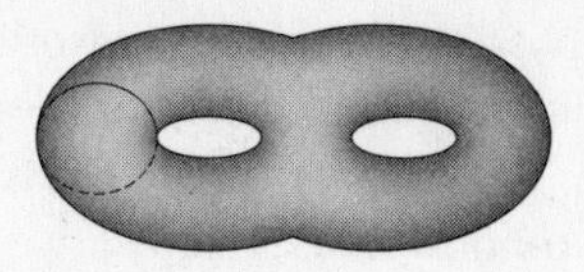

FIGURE 9. *Two-holed torus*

what type of manifold we were living on? No explorer would encounter any holes—everything connects up.

To see what an atlas of this world might look like, we can cut it up into regions that could be easily mapped. In fact, we can make a single world map by matching up the regional maps. Imagine cutting the world along the four curves starting at a single point, as shown in figure 10. To keep track of what needs to be matched up to what, let's label the curves *A*, *B*, *C*, and *D*, as shown below, and cut in the directions shown. Starting with A and C, we can imagine unrolling the surface onto a flat piece of paper shaped like a figure eight on its side. Cutting along the remaining two curves (which are now line segments) gives a shape that we can deform into an octagon, and even a rectangle. This gives a world map with edges that need to be matched up as indicated.

Can we list all possible two-dimensional manifolds? After Magellan's expedition returned, but before anyone had explored the poles, could we have

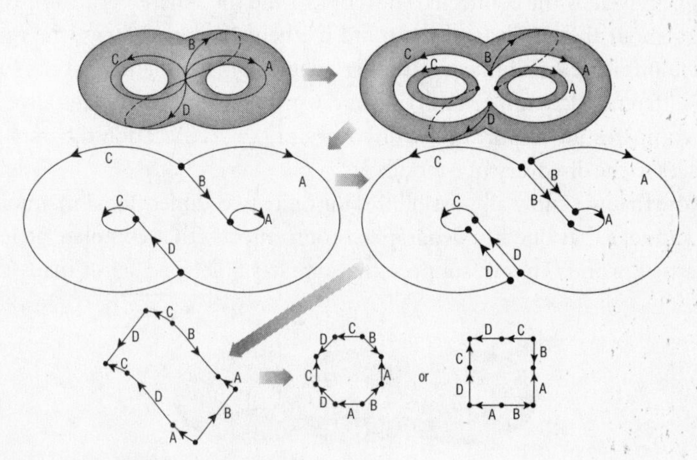

FIGURE 10. *Cutting a two-holed torus*

been ready with a set of possible shapes for our world? The answer, together with its proof, is one of the great achievements of the nineteenth century.

The answer, that is the list of all possible shapes, is deceptively simple. We might be inhabiting a two-holed torus. Likewise, we could consider a three-holed torus (see figure 11), or four-holed a torus, or for that matter, tori with any number of holes. These are all different, and they exhaust the possibilities for orientable two-dimensional manifolds. All of these are possible shapes for a world—if our world were in the shape of any one of them, and we were at any point of that world, we could map the region around us.

A nice way to think of these different manifolds is in terms the notion of a connected sum of two manifolds. Suppose that we have two two-dimensional manifolds and that we cut a disk out of each.

The boundary of a disk is a circle. We can make a new two-dimensional manifold without boundary by sticking the resulting manifolds-with-boundary together along their boundaries, identifying the boundary circles of each. This manifold is called the *connected sum* of the two original manifolds. Figure 13 illustrates that the two-holed torus is a connected sum of two tori. A three-holed torus is a connected sum of a two-holed torus and a regular torus, and therefore a connected sum of three tori. A four holed-torus is a connected sum of four tori, and so on.[23]

Thus, a complete list of finite two-dimensional manifolds that have no boundary, and that are orientable (that is, that could be the surface of something), is given by the connected sums of tori and the sphere.[24] The more one thinks about this result, and the more mathematics one knows, the more miraculous it seems. Surfaces arise everywhere in mathematics and can look very different. Vladimir Arnold, one of the greatest mathematicians alive today, compares the import of the discovery of the classification theorem for surfaces to the discovery of America.[25]

Now that we know all possibilities for finite two-dimensional manifolds, let's suppose that one is a geographer confronted with a complete atlas of some new world. That is, suppose that one has a detailed set of maps that

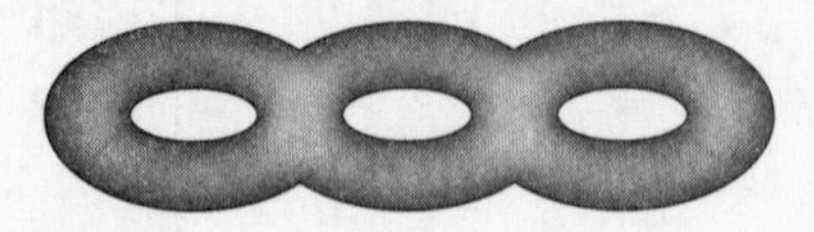

FIGURE 11. *Three-holed torus*

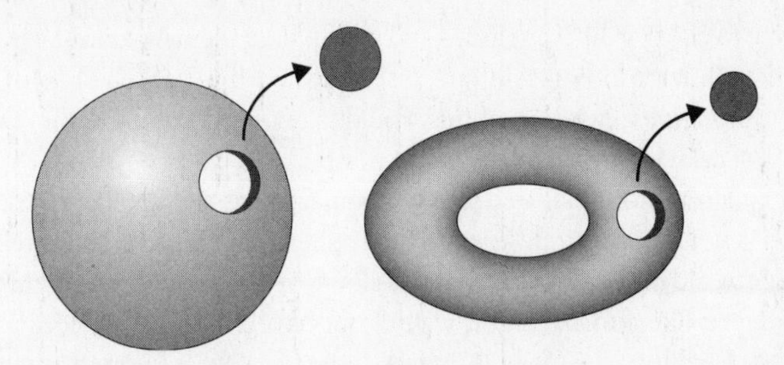

FIGURE 12. *Cutting a disc out of a sphere (left) and torus (right)*

cover every region, but that one doesn't know the shape of the world. Could one figure out which two-dimensional manifold it is? One could try to assemble all the maps together, but this can be quite difficult—imagine trying to assemble maps if the world were a two-holed torus, or worse!

This question gives rise to some lovely mathematics. The central idea is to consider *closed paths* or *loops* on the surface. These are paths that begin and end at the same point. Assuming one lives on the surface, one can think of a loop as a round trip: It is the path one traces out when one takes a journey on the surface and returns where one started. Think of it as the curve that a string would make it if it were trailed out behind as one took the journey. Mathematicians typically introduce relations among loops, called *homologies,* that

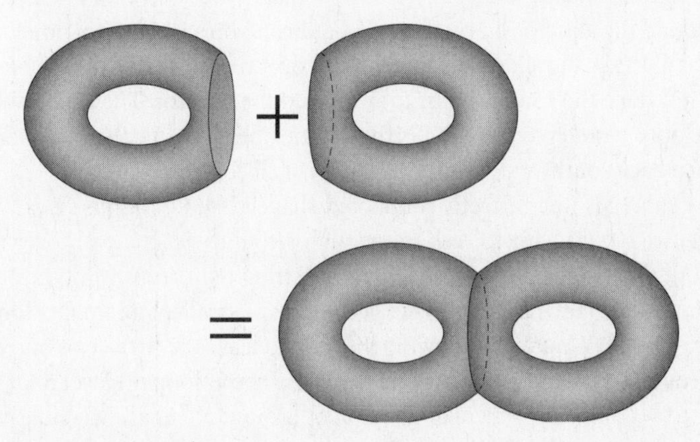

FIGURE 13. *Connected sum*

allow one to classify different loops and manipulate them like numbers. Manifolds that differ topologically are distinguished by the behavior of their loops. Although explanation of this would take us too far afield, there is one instance of critical importance to us.

To explain it, note that one says colloquially that a torus has a hole, and that this is what makes it different than a sphere. However, if we were living on a huge world (again, let's say that it is cloud-covered so that we couldn't see off it) shaped like a torus, and if we did not have access to a spacecraft, we would not be able to see the hole. In fact, if the torus were knotted as in figure 8, we might even have trouble explaining what we meant by a hole.

There is, however, a way to distinguish between living on a torus from a sphere. On a torus (or any connected sum of tori) are some loops that are essentially different from any on a sphere. To see this, imagine living on a manifold and taking a round trip from some point. Imagine laying out string, say a very thin fishing line, as we go along, making sure to firmly attach the string to a stake at the starting point. If we pause at some intermediate point of this journey and try to reel in the string from where we are, it will pull taut held at its other end by its stake. The string will trace a direct path from the point we stopped to the point where we started (and measuring the length of the string would give you the exact distance between the starting and stopping points). Let's continue now to the end of the round trip.

We are now back where we started and have laid out a gigantic loop of string. Reel it in. Can you reel it all the way in, remembering that the starting point is fixed? Whether on a sphere or on a plane, the answer is yes. Even a loop around the equator on earth could be shrunk into its starting point. As we reel it in the string slides, say, northward, over the surface of the sphere—it does not trace the route we just took around the equator. The smaller and smaller loops would go over the North Atlantic, then Canada, then the North Pole, then back south, and so on, until finally, it all comes down to us.

If the sphere is not perfectly round, but shaped like Columbus's pear, for instance, we would have to reel out enough string to get the path over the bump, but then we would be able to pull the string tight and shrink the path to a point. As we reel the string in, we get a series of smaller and smaller loops *on* the surface.[26] (We are not allowing the loop to leave the surface or burrow underground!) However, on a torus, there are some loops that cannot be shrunk to a point. In particular, if we went around a "hole," we could not shrink the loop beyond that hole's diameter. The same is true on a connected

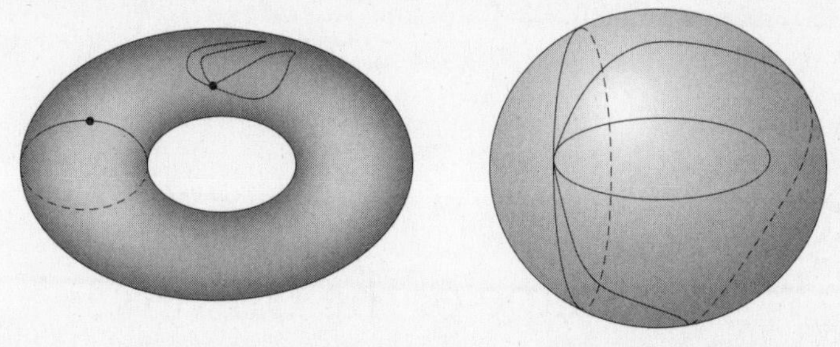

FIGURE 14. The leftmost loop on the torus above cannot be shrunk to a point staying on the surface; the two other loops can be. Every loop on a sphere can be shrunk to a point.

sum of tori. Figure 14 shows a loop on the torus that cannot be shrunk to a point, as well as some loops that can.

If every loop on a manifold can be shrunk to a point, we say that the manifold is *simply connected*. Because of the classification theorem for two-dimensional manifolds, we can see that the sphere is the only simply connected two-dimensional manifold. Although it is difficult, a geographer working with an atlas can determine whether the manifold it depicts is simply connected and hence whether or not the world is a sphere.

We have not yet defined a *three-dimensional manifold*. Just as two-dimensional manifolds model worlds, three-dimensional manifolds are the mathematical objects that model universes. There is a particularly nice three-dimensional manifold called the *three-dimensional sphere*, which is finite, does not have a boundary, and has the property that every loop can be shrunk to a point. The Poincaré conjecture states that this is the *only* finite simply connected three-dimensional manifold. What exactly does this mean, and what is the three-dimensional sphere?

4

The Shape of the Universe

Just as in the case of our Earth, an atlas of the universe would be a collection of maps. But a map of a region of the universe would not be a rectangular piece of paper. Rather it would look like a solid glass box (think of an aquarium or transparent shoebox) filled with clear liquid crystals in which spots would be lit up to correspond to the positions of planets and stars. In the box-map that contained our solar system, if we looked at a distance corresponding to 431 light years straight up from the Earth, you would see a point corresponding to Polaris, the North Star. Looking away from the Sun in various directions in the plane of the Earth's orbit, we would see the other planets. South of the plane of the equator in one direction, at a distance corresponding to a little more than 4 light years, would be our closest neighbors, Proxima Centauri and the double star Alpha Centauri. Depending on the scale of the box-map, out in another direction, also south of the equator, there would be the galactic center, with its massive black hole, at a distance corresponding to 25,000 light years. Further away in a somewhat different direction would be Andromeda, the spiral galaxy nearest to us, at a distance corresponding to 2.9 million light years away.

An atlas of the universe would be a collection of such transparent shoeboxes with every region mapped in at least one box. If, as seems likely, the universe does not go on forever, the number of boxes necessary to make up a full atlas would be finite. However, we can't "see" the universe put together as a whole. If we had a complete atlas of the universe, so that every part of it was mapped, we could try to assemble our clear shoebox maps. Yet, just as

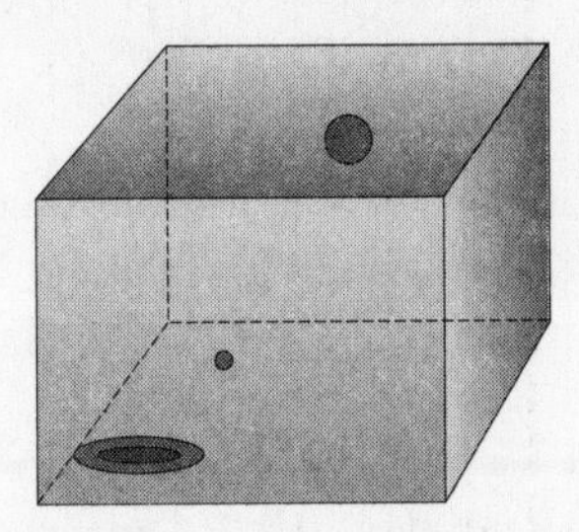

FIGURE 15. A map of a region containing a small spiral galaxy at lower left front.

there is not enough room on the plane to put together maps of our world into a globe, there is not enough room in ordinary space to fit all the shoebox maps together nicely. We have trouble picturing the shape of a universe as a whole.

Moreover, it is not possible to get outside the universe. This is an important difference between the Earth and the universe. A rocket can leave the surface of the Earth, and we can look at the Earth from outside it. Since we see in three dimensions and since the surface of the Earth is two-dimensional, we can see our planet bending in a third direction and visualize its whole shape easily. Even if we could get outside the universe in an attempt to see what shape it had, since the universe is three-dimensional, we would need to be able to see in at least four dimensions to visualize the universe as a whole.

As we shall see, this does not mean that the universe does not have a shape. It also does not mean that the universe cannot curve. It could have many different shapes; and it is virtually certain that the universe, just like the surface of the Earth, curves differently in different places.

Although the universe is immensely larger than the Earth, the study of the universe has some distinct advantages unavailable to the heirs of Pythagoras as they puzzled about the surface of our planet. Unlike the Earth where our vision is cut off by a horizon so cramped that we have to travel to make a map that covers a reasonably large area, we can see for a long way into the universe using telescopes. We have also got quite good at measuring distances to the various objects we see in the skies. Thus, we can construct a map of a rather large region of the universe without actually traveling off the earth. Our mathematics is much further advanced than it was in Columbus's time, and there are many powerful mathematical tools that can be brought to bear on the question of the shape of the universe.

FINITE UNIVERSES

Although the Greeks believed that the earth was spherical, many of them felt the universe extended infinitely. Here, for example, is Archytas, the very accomplished Pythagorean mathematician and friend of Plato, on the subject.

> If I were at the outside, say at the heaven of the fixed stars, could I stretch my hand or my stick outward or not? To suppose that I could not is absurd: and if I can stretch it out, that which is outside must be either body or space (it makes no difference which it is as we shall see). We may then in the same way get to the outside of that again, and so on, asking on arrival at each new limit the same question; and if there is always a new place to which the stick may be held out, this clearly involves extension without limit. If now what so extends is body, the proposition is proved; but even if it is space, then, since space is that in which body is or can be, and in the case of eternal things we must treat that which potentially is as being, it follows equally that there must be body and space extending without limit.[27]

Archytas is arguing that there is no boundary to the universe—wherever we are in the universe, he argues, when we look in the sky, apart from whatever bodies it contains, it will look roughly the same. We won't see an edge—wherever we are, we can stick a hand or stick out. Since there is no boundary, he concludes erroneously that the universe must be infinite. To see why his argument is flawed, note that we could repeat the same argument on the face of the Earth. When we stand outdoors, anyplace on our planet, we can stretch out a hand or stick horizontally and no barrier prevents us from extending our own body in this fashion. There is no boundary, no edge. If Archytas's conclusion were correct, the Earth would extend infinitely in all directions. But it does not. It is a sphere. (It could even have been a torus.) To say that the universe has no boundary is not to say that it goes on infinitely, just as to say that the Earth has no edge is not to say that it goes on forever.

It may be that the universe goes on forever, but it seems very unlikely. Space and matter are intimately related, and the assertion that the universe has an infinite amount of matter causes serious theoretical problems. The universe could also have a boundary of some kind, but this is a bit like assuming that the world is a disk with an edge that one could fall off. Few scientists with mathematical training seriously believe this, either.

Just as there are many possible shapes that earth might have had (it could have been a sphere, a torus, a connected sum of two tori, and so on), there are different shapes the universe might have. Infinitely many. And unlike the situation for surfaces, we do not even have a classification of the possible shapes that the universe could have. There are far too many.

Regarding the size and shape of the universe, we are almost in *precisely* the same position that Columbus was in 1492. Just as there was no complete atlas of the Earth in Columbus's time, there is no complete atlas of the universe today. If we left the Earth on a very fast spaceship, headed out in a fixed direction (and took care not bump into planets or stars), after a very long time, most cosmologists and mathematicians believe, we would come back close to where we started. Of course, the distances are much further and the speed of light seems to put an upper limit on how fast we can physically travel, but the statement that travel away from the Earth in the same direction would return us to our starting point is no less meaningful or paradoxical than Erastosthenes' statement, or John Mandeville's seventeen hundred years later, that sailing due west from Spain would bring us back to where we started. And, just as in Columbus's time, there are competing estimates for the distance we would go before returning.

THREE-MANIFOLDS

Since we can't see in more than three dimensions and since we can't get outside the universe, we have trouble picturing the shape of the universe as a whole. Here, even more so than was the case for surfaces, we need to be precise about what we are talking about. It will be remembered from the last chapter that a two-dimensional manifold is a mathematical object that shares a key property with the surface of our earth. Namely, all regions can mapped onto on a piece of paper. Thus, if one imagined oneself as very tiny and at any point of a two-dimensional manifold, the area around one would seem to extend away infinitely, making it look as if one were living on a plane. A sphere and a torus are examples of particular two-dimensional manifolds, and we talk of possible shapes of the Earth in terms of different two-dimensional manifolds. Two-dimensional manifolds are the mathematical objects that model possible worlds.

The corresponding mathematical object that models our universe is a *three-dimensional manifold,* or *three-manifold.* It is a set in which every point belongs to a region that can be mapped onto the points inside a clear aquarium or shoebox. In other words, the region around any point looks like space rather than a plane. As before, we say that an *atlas* is a collection of maps that is *complete* in the sense that every point belongs to some region that is covered by one of the maps. A three-manifold is the object that is covered by all the maps in an atlas.

Somewhat simpler to picture than three-manifolds are *three-manifolds-with-boundary*. An example is a solid ball: think of the Earth, but now include not just its surface, but everything inside it. If one were inside the earth, one could map all the points around one by having them correspond to points inside an aquarium. The boundary, that is, the surface, is a two-dimensional sphere. Similarly, a solid torus is also a three-manifold-with-boundary. Think of a doughnut or an anchor ring: The boundary is again a two-dimensional manifold, but this time a torus. If one is inside a solid torus, not on the surface, one can again map the points around one by making them correspond to points inside an aquarium.

If you have two three-manifolds-with-boundary and their boundaries (which are two-dimensional manifolds) are homeomorphic to one another (so, for example, if each boundary were a sphere), then one can stick them together along their boundaries by deeming that points that correspond under a particular homeomorphism of the two boundaries are the same (points not on either boundary stay distinct). The resulting mathematical object is a three-manifold without boundary. This is because the former boundary points stop being boundary points. At a former boundary point, the region on one side of the boundary belongs to one manifold, the region on the other side to the other manifold, and one can pass from one to the other. Put differently, at a former boundary point, the region around one would now be filled up with points of one manifold or the other, and one could map it using a solid shoebox.

Just as for surfaces, two three-manifolds are the *same topologically* if the points of one can be put into continuous one-to-one correspondence with the points of the other. (Remember, *continuous* is a technical term that captures the notion that nearby points correspond to nearby points.[28]) Two three-manifolds that are the same topologically are also said to be *homeomorphic,* and the one-to-one correspondence that establishes that they are the same is called a *homeomorphism.*

A three-dimensional manifold is called *compact* or *finite* if there is an atlas of it that is finite. The three-dimensional analogue of the plane envisaged by Euclid[29] that goes on forever is a three-dimensional manifold that is not finite. The very simplest finite three-manifold is the *three-dimensional sphere,* or *three-sphere.*

THE THREE-SPHERE

Consider two solid balls (figure 16). Each is bounded by a two-dimensional sphere *and* includes the region inside. Now imagine that these two balls are glued together along their boundary. That is, we declare that points on the boundary spheres that correspond to one another are in fact the same point. We can't physically glue the boundary spheres together in three-space, but that is no matter. We can imagine exactly what living in such a universe would be like. When we travel, or look, across the sphere originally the boundary of one ball, we continue on right to the next. We don't see any break at all. It is just as if we crossed the equator on our two-dimensional earth—we can simply see across it without there being an actual line there.

Remember, at any point in this universe, we can make a map of the region around ourselves using the inside of a shoebox. Another way of thinking about this construction is to imagine both solid balls as maps that are like the box-maps with which we started, but which are transparent solid balls instead of transparent solid shoeboxes. Imagine each mapping half of a universe. That is, there are images of galaxies and nebulae mapped inside each, but each half is completely different except that the objects on the boundary correspond.

Notice that the same construction applied to two disks (that is, two circles

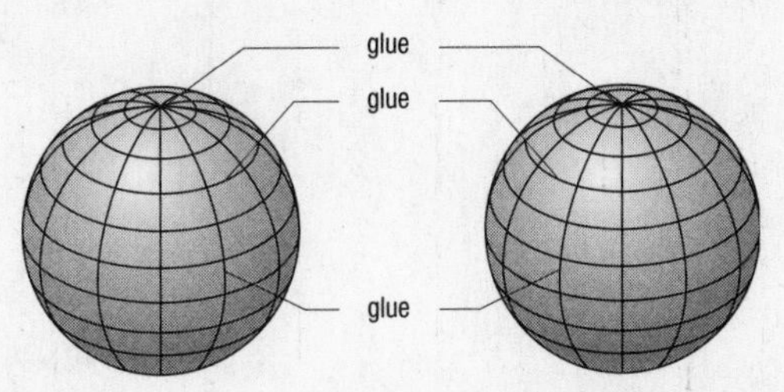

FIGURE 16. *Matching up corresponding points on the boundary of each solid ball gives a three-sphere.*

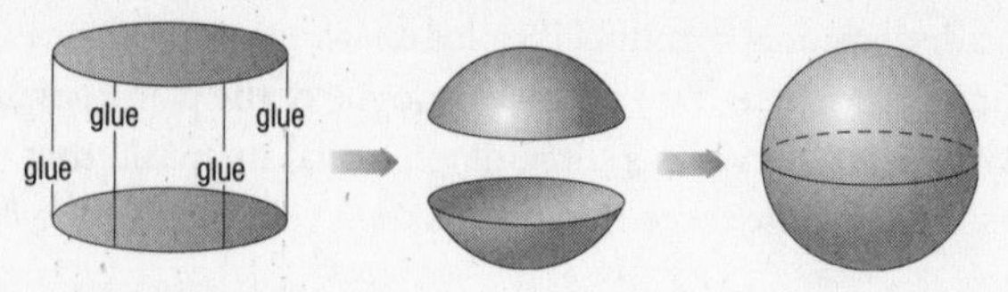

FIGURE 17. *Connecting the boundary circles of two (two-dimensional) disks gives a two-dimensional sphere.*

in two-space together with the region in the plane inside the circles) gives us the two-dimensional sphere. These disks form two hemispheres on the two-dimensional sphere (see figure 17). The two-dimensional sphere is obtained by gluing the boundaries of the two disks together. Here we have an advantage, because we can pop the disks out of the plane and imagine them in three-space, where we have no trouble visualizing the gluing.

Here, again, it is useful to think of the disks as maps. Indeed, a common representation of our earth is achieved by two maps that are not rectangular, but as disks bounded by circles. Figure 18 shows one such representation where one disk shows the hemisphere to which the Americas belong and the other disk shows the opposite side with Europe, Asia, Africa, and Australia.

We have more difficulty imagining the three-sphere as a whole because we don't have an extra dimension to get outside it. In the case of the three-sphere, the hemispheres are not two-dimensional disks with boundaries; they are the two solid balls, and the common boundary is not a circle, but the two-dimensional sphere.

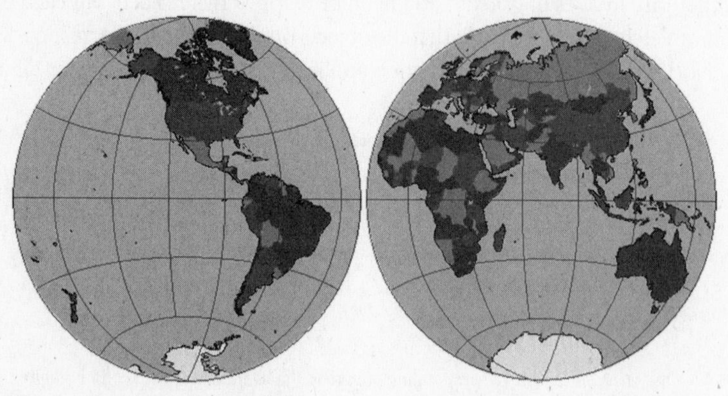

FIGURE 18. *A hemispheric world map*

A number of scholars have argued convincingly that the universe imagined in *The Divine Comedy* by the great Italian poet and writer Dante Alighieri (1261–1321) is a three-sphere (although he did not, of course, call it that).[30] In "Paradiso" he climbs up from hell at the center of the Earth to the surface of the Earth through the concentric spherical shells in which the various planets live, up past the spherical shell containing the fixed stars reaching the Ninth Heaven, or Primum Mobile. There at the top of the Primum Mobile he looks down with his love, Beatrice, over the half of the universe he has traversed and out to the heavenly half of the universe, consisting of a concentric spherical shells where the angels, then archangels, and higher-order angels live. The two-dimensional sphere at the edge of the spherical shell that is the Primum Mobile is the equator from which he and Beatrice survey the universe. Earth (and hell at the center of the Earth) are at one pole. At the other pole is the realm of the Seraphim.

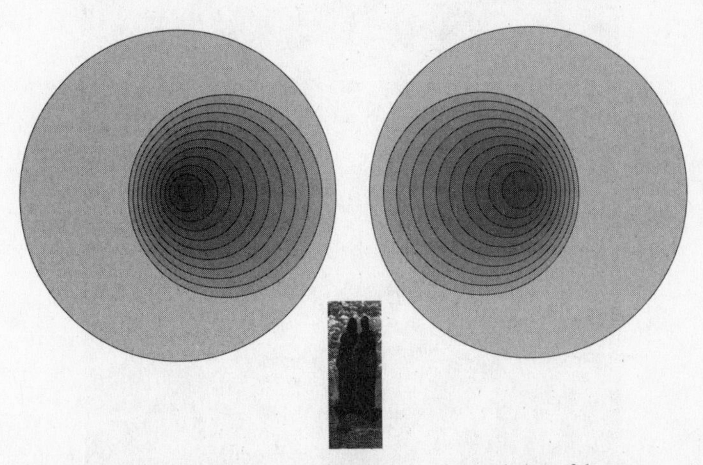

FIGURE 19. *Dante and Beatrice (center) staring into the two halves of the universe. At left is a cutaway showing the concentric spherical shells of the visible hemi-universe: The outermost spherical is the Primum Mobile (at whose edge Dante and his muse are standing), then the shell of fixed stars, followed by those of Saturn, Jupiter, Mars, Sun, Venus, Mercury, the Moon, and finally the Earth at the center. On the right is Empyrean, the angelic universe filled out by the spherical shells of the angels, archangels, principalities, powers, virtues, dominions, thrones, cherubim, and the sphere, with the sphere of the seraphim at the center.*

In figure 19, each of these halves represents a solid ball through which we have cut a conical wedge so that we can see into it. The earth is in the center of the left-hand ball and the realm of the Seraphim in the center of the right-hand one. There is a lovely print by Gustav Doré of Dante and Beatrice (figure 20) gazing up at the heavenly part of the universe, with the realm of the Seraphim far away at the center. Unfortunately, the sketch does not get it right—Doré has drawn rings, instead of concentric spherical shells, of angels. It becomes right if, instead, we imagine Dante and Beatrice looking down a conical opening to the bright heaven that is at the center of the solid sphere into which they are looking, and if we interpret the rings of angels as the circular edges of the opening cuts in the spherical shells. The spherical shells then wrap away from the viewer back behind the shining center.

The three-sphere is finite. We could cover with a finite number of box-

FIGURE 20. *Doré's sketch of Dante and Beatrice looking into the heavenly hemisphere.*

maps all regions mapped by the two solid balls that we glued together. (Of course, some maps would have to cover territory in both balls.) Moreover, there are three-spheres of any size. We can take the solid balls we glue together to be as large or small as you like.

Finally, and this is a little more subtle, every loop in the three-sphere can be shrunk to a point. To see this, imagine that we are living in the three-sphere represented in figure 16 and that take a round trip, stretching out a fishing line as we go. We might, for instance, begin at some point far inside the left-ball, travel outward until we reach the boundary, then enter the right-hand ball, continue until we got to the other side of it, at which point we would be back in the left-hand ball and would continue back to the point at which we started, where we would reel the line in. Even though the beginning point of string is attached, nothing stops us from gathering back the string that has been reeled out, making smaller and smaller loops beginning and ending at our starting point. Eventually the loops are entirely in the left-hand solid ball and we can keep pulling line in until finally the loop shrinks to the starting point. Hence the three-sphere, like the two-dimensional sphere, is simply connected: any loop can be shrunk to a point.

MORE COMPACT THREE-MANIFOLDS

There are many compact three-manifolds different from the three-sphere. Consider, for example, a solid spherical shell consisting of the region in space between two concentric spheres. Using food imagery, think of the "meat" of an avocado: the edible region outside the inner seed and inside the outer skin. This is a three-manifold-with-boundary, where the boundary consists of two two-dimensional spheres. If we imagine the inside sphere attached to the outside one by matching every point on the inside sphere with the point closest to it on the outside sphere (in other words, match the points on the inside boundary sphere with the point on the outside boundary sphere radially outward from it), we get a nice three-dimensional manifold without boundary.

To imagine living in a universe with this shape, imagine being very tiny and floating inside the spherical shell that is very big, and subject to the proviso that, if we travel outward, piercing the outer two-sphere, we don't leave the shell, but immediately reenter via the inner two-sphere. We would not, of course, see outside, because the inside spherical boundary is actually the same as the outside spherical boundary (see figure 21). Nor could we see these "boundary" spheres: looking out the outer sphere results in seeing in the inner sphere. In fact, these "boundary" spheres are an artifact of the construction: From the point of view of people dwelling in this manifold, they would not be there. At any point, if we looked around, it would appear as if we were floating in a region in Euclidean three-space, and in fact a map could be drawn of the region near us. Since the set, although very large, is not infinite, we could construct an atlas of finitely many shoebox-type maps that would map every region.

The difference between living in this manifold and the three-sphere only becomes apparent globally. If we were to travel in a direction corresponding to radially outward, we would come back to where we started. So we would traverse a closed loop. If we imagine laying out string along this path, there is no way we could pull it into a point. This loop of string can never be shrunk smaller than the distance between the original inside and outside two-spheres that were identified to make the manifold. So, this manifold has to be different from the three-sphere.

In other respects, however, this manifold is like the three-sphere. It has no walls, but it is finite. Going out in any direction will bring us back near to where we started. It is a possible universe. One might object that this manifold can't really exist because we cannot physically connect up the inside and outside spherical boundaries. But this is to miss the point. Attaching the inside and outside spherical boundaries in our imagination is only our way of visualizing this manifold. The manifold is the primary object, and there can

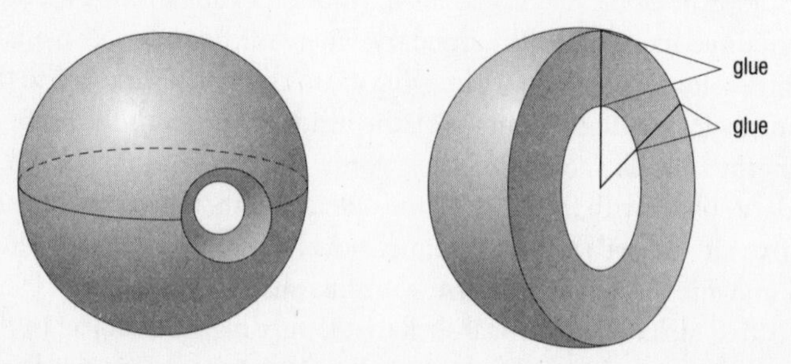

FIGURE 21. *Joining each point of the inside boundary sphere to the corresponding point radially outward on the outside boundary sphere gives a three-manifold-without-boundary.*

be no question that the mathematical object exists. We don't know whether our universe is shaped like this. Like the three-sphere, this manifold is a potential shape that our universe might have.

Another example of a three-manifold is obtained by considering a solid rectangle (or aquarium). Its boundary consists of six faces, consisting of three pairs of parallel rectangles. Suppose we connect opposite faces by deeming that points that are directly across from one another are the same. This means that if we go out the right side, we come back in on the left; out the top, we come back in the bottom; and out the back, we come in the front. Again, it is important to emphasize that there are no boundaries—if we fly around in this manifold, we never run into a wall and—yet we never leave it. This may sound far-fetched. It is not. In the last chapter, we saw that it was easy to imagine a world with a world map such that the places represented on the left edge are the same as those on the right edge, and those on the top edge the same as those on the bottom edge. Such a world is a two-dimensional torus. The three-dimensional analogue formed by connecting opposite faces of a rectangular solid is called a *three-dimensional torus,* or *three-torus* for short.

The representation of the two-dimensional torus as a rectangle with its opposite sides attached gives us another way to think about the three-torus. Let's consider the original rectangular solid as being composed of infinitely many rectangular sheets of paper, say, stacked front to back. Matching the top face of the solid with the bottom, and the right face with the left, automatically matches the top edge of each sheet of paper with the bottom edge, and the right edge with the left edge. In other words, we wind up with a solid that is "fibered" by two-dimensional tori stacked against one another. We get a *toroidal shell*: a solid region between two two-dimensional tori, one inside the other, that is filled out by concentric tori of increasing size. The boundary consists of the two two-dimensional tori (corresponding to the sheet of paper in the front of the stack and the sheet of paper at the back). See figure 22. To get a three-torus, we connect the two two-dimensional tori that make up the boundary. If we were living in this manifold, we could imagine ourselves living in the toroidal shell, where the outside torus is the same as the inside torus, so we can't see them, and passing through one results in coming back in the other.

This manifold again has closed paths that cannot be shrunk to a point. In fact, if we start out from a point in the middle of the shell and drive straight outward, going out the outer torus and, hence, back in the inside torus, we get

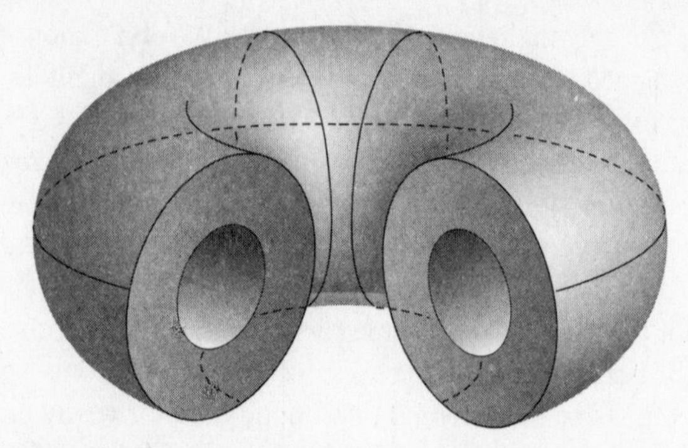

FIGURE 22. *The three-torus is obtained by connecting corresponding points on the inside and outside tori of a toroidal shell.*

back to where we started. See figure 22. We can't shrink this path to a point. So this manifold can't be homeomorphic to the three-sphere, any loop in which can be shrunk to a point. It is also not homeomorphic to the manifold we constructed by attaching the inside and outside boundaries of a spherical shell.[31]

Incidentally, the concept of making a world map that gets in as much as possible also applies to the universe. Imagine having mapped our entire universe so that we have hundreds of clear glass boxes mapping the stars and galaxies in different regions of space. We would, naturally, rescale so they were all to the same scale, and try to fit as many as possible together to create a big solid map of the universe. But just as in the case of the world, there is only so far we can go without having the faces of some of the blocks on an edge connecting with the faces of some other blocks on another edge. Just as we cannot connect the points at the edges of our world map by staying in a two-dimensional plane, we cannot connect the faces at the outside boundary of our universe map/tank by staying in our space. This should not dismay us. The two-dimensional plane and three-dimensional space of high school geometry that go on forever are mathematical objects that are just as much conceptual constructions as the manifolds we are talking about.

THE POINCARÉ CONJECTURE

We have seen a few examples of compact three-manifolds: the three-torus, the three-sphere, and the manifold we get from connecting the inside and outside boundaries of a spherical shell. There are many more, infinitely many. There are also infinitely many two-dimensional manifolds, but they have all been classified. Not so three-dimensional manifolds. There is a far richer variety of them, and no one has ever come close to classifying them all. Human ingenuity is close to boundless, and a great deal of it has been expended constructing different three-manifolds. We can construct three-manifolds by joining opposite faces of regular solids in different ways. Given a three-manifold, we can cut a solid torus out of it and glue it back in differently to get another, often different, three-manifold. Given two three-manifolds, we can cut a solid ball out of each and glue the resulting two manifolds-with-boundary together by connecting the points of the two-dimensional spheres that bounded the solid balls we removed. Any of these three-manifolds is a potential shape that the universe might have. We have an embarrassment of riches. Is there any way to make sense of it all?

Over the last century, many individuals have devoted their life's work to furthering our understanding of three-manifolds. But, maddeningly, the simplest question was the one that eluded all efforts at arriving at an answer: Among all those three-manifolds, is there any one that is different from the three-sphere and that has the property that every path can be shrunk to a point? If there is no such manifold, then we could say for sure whether our universe is a three-sphere by using a complete atlas to check whether every closed loop could be shrunk to a point. The Poincaré conjecture states that there is no such manifold. More formally, and phrased more positively, the Poincaré conjecture is the assertion that any compact three-manifold on which any closed path can be shrunk to a point, is the same topologically as (that is, homeomorphic to) the three-sphere.

It is the simplest question we could ask about the potential shape of our universe. Yet it has been a heartbreaker and, for some, an obsession and career-killer. This is what drew the crowds at MIT and Stony Brook. This is the conjecture that Perelman settled. And it is this conjecture that carries a million-dollar award.

5

Euclid's Geometry

If our universe doesn't go on forever, and if it has no walls, doesn't that mean that the universe must be bent or curved somehow? To rephrase in terms of the language of the last chapter, shouldn't any compact three-dimensional manifold without boundary curve back on itself? How else can we explain going off in one direction and coming back to where we started? It is one thing to say that the surface of our world is two-dimensional in the sense that we can map regions of it on flat pieces of paper, but doesn't it need a third dimension in which to curve? And, shouldn't the same be true of our three-dimensional universe? If it is curved, whatever that means, doesn't it need a direction in which to curve? But if our universe includes everything, how can there be something else to curve into? Come to think of it, what do we mean by "curving"? Do these questions mean anything, or are they misguided wordplay with poorly defined terms?

These questions are meaningful, and turn out to be critical to the Poincaré conjecture and its proof. They also illustrate the reason that mathematicians insist on absolute rigor. Anytime we communicate with another person, we invoke years of shared experience. We know that a glass will not fall through a table, that buildings have insides accessible by doors, that one can be right-handed or left-handed. We know what it is to be in love or to feel pain, and we don't need precise definitions to communicate. The objects of mathematics lie outside common experience, however. If one doesn't define these objects carefully, one cannot manipulate them meaningfully or talk to others about them.

Artists and humanists embrace complexity and ambiguity. Mathematicians, in contrast, work by obsessively defining terms and stripping off extraneous meaning. The almost neurotic insistence that every term be rigorously defined, and every statement proved, ultimately frees one to imagine and talk about the unimaginable. Most people, traumatized by school experiences of mathematics, know all too well that mathematics is the most meticulous and demanding of disciplines, but few get to see that it is also the most liberating and imaginative of all human activities. Absolute precision buys the freedom to dream meaningfully.

But absolute precision comes at a price. Terms need very careful definitions, and every statement, even the seemingly obvious, must be proved. What seems obvious can be frighteningly difficult to prove—sometimes, it even turns out to be wrong. Seemingly tiny exceptions matter, details can overwhelm, and progress can be unbearably slow. Mathematics is the only field of human endeavor where it is possible to know something with absolute certainty, but the hard work of slogging through morasses of possible definitions and formulations too often foreclose the dreamy vistas it affords to all but a driven few. Nothing illustrates the tension between precision and dreaminess better than Euclid's *Elements*, the famous treatise on geometry that is essential to our story, and to which we now turn.

THE *ELEMENTS*

Euclid's *Elements* dates to the reign of Ptolemy Soter (the first Ptolemy) around 300 BCE in Alexandria. From the beginning, it was a sensation. The *Elements* codified the mathematics developed from the times of Thales and Pythagoras through Plato and Archimedes. It reinterpreted millennia-old Babylonian and Egyptian mathematics within a distinctively Greek framework.

Sadly, we know almost nothing about Euclid (c. 325–c. 265 BCE).[32] We know even less about him than we do about Pythagoras, and what little we do know has been hotly contested by scholars. Euclid wrote at least ten books, only half of which have survived. A number of mutually consistent indications suggest that he lived after Aristotle and before Archimedes. He was one of the first mathematicians at the great library of Alexandria and there had gathered a group of talented mathematicians about him. Legends about him abound, many as (possibly apocryphal) insertions in other mathematicians' works. One tells that Ptolemy asked Euclid for a quick way to master geometry and received the reply, "There is no royal road to geometry." Another tells of a student who, after encountering the first proposition in the *Elements*, asked Euclid what practical use studying geometry could have. The mathematician allegedly turned to his slave and replied dismissively, "Slave, give this boy a threepence, since he must make gain of what he learns."

The *Elements* contains thirteen books (chapters). Books 1 to 6 deal with plane geometry, 11 to 13 with solid geometry, and 7 to 10 with number theory. Everything is worked out from first principles. Book 1 starts with twenty-three definitions, five common notions, and five postulates. The definitions name the basic objects and concepts that Euclid will consider. The *common notions* are commonly accepted rules about reasoning and relationships that he makes explicit. The *postulates*, or axioms, are assertions about the objects under consideration that are assumed to be true without proof. Nowadays, we would treat the common notions as axioms as well. The definitions, common notions, and postulates are taken as the starting points from which further assertions, called *propositions*, are proved by strict logical rules. An especially significant proposition is called a *theorem*, a proposition whose main purpose is to prove a theorem is called a *lemma*, and a proposition which follows especially easily from a theorem is called a *corollary*. A *proof* of a proposition is an ordered, precise deductive argument in which each assertion is an axiom or previously proved proposition, or else follows from such by formal rules of logic. It begins with axioms and known propositions, and ends with the statement that is to be proved.

Here, for instance, are some of the definitions in book 1.[33]

1. A *point* is that which has no part.
2. A *line* is a breadthless length. . . .
8. A *plane angle* is the inclination to one another of two lines in a plane which meet one another and do not lie in a straight line.
9. And when the lines containing the angle are straight, the angle is called *rectilineal*.
10. When a straight line set upon a straight line makes the adjacent angles equal to one another, each of the equal angles is called *right*, and the straight line standing on the other is called *perpendicular* to that on which it stands. . . .

23. *Parallel* straight lines are straight lines which being in the same plane and being produced indefinitely in both directions, do not meet one another in either direction.

Here are the five common notions.

1. Things which are equal to the same thing are equal to one another.
2. If equals be added to equals, the wholes are equal.
3. If equals be subtracted from equals, the remainders are equal.
4. Things which coincide with one another are equal to one another.
5. The whole is greater than the part.

And, here are the five postulates.

1. There is a straight line from any point to any point.
2. A finite straight line can be produced in any straight line.
3. There is a circle with any center and any radius.
4. All right angles are equal to one another.
5. If a straight line falling on two straight lines makes the interior angles on the same side less than two right angles, the two straight lines, if produced indefinitely, meet on that side on which the angles are less than the two right angles.

Euclid then goes on to deduce all of the common truths of plane geometry, each using only the definitions, common notions, postulates and previously established propositions. So, for example, proposition 1 states that given any line segment, we can create an equilateral triangle one of whose sides is that line segment. Proposition 4 states that if two sides of one triangle and the angle they enclose are equal to two sides and the enclosed angle of another triangle, then the two triangles are equal. Most high school texts refer to this as the SAS (side-angle-side) property. Proposition 5 tells us that the angles at the base of an isoceles triangle are equal (the definition of *isoceles* requires only that two sides be equal; one must then prove that the angles these sides make with the third side are equal). Proposition 32 states that the angles of a triangle add up to a rectilineal angle (that is, the sum of the angles of any triangle is 180 degrees) and proposition 37 gives the formula for the area of a triangle. Proposition 47 is the Pythagorean theorem: The sum of the areas of squares erected on two sides of a triangle that meet at a right angle (that is, at 90 degrees) equals the area of the square on the side opposite the right angle; and proposition 48 its converse (if the area of the square erected on the long side of a triangle equals the sum of the areas of the squares on the other sides, the angle opposite the long side is a right angle). Two thousand years of geometry in a few short, compelling pages!

Earlier texts had appeared before Euclid's with the same emphasis on deduction. But Euclid got it right. His choice of axioms, his arrangement of propositions, and the sheer coverage were brilliant and rendered other texts obsolete. Plato, Aristotle, and other Greek philosophers were enormously interested in mathematics, and the two centuries from the times of Thales and Pythagoras to the establishment of Alexandria saw much discussion of geometry, along with arguments about what followed from what, and learned exchanges over appropriate first principles. Euclid brought order to a large collection of scattered proofs and discussions. The findings of over two hundred years of Greek geometry and number theory, and fifteen hundred years before that of Babylonian mathematics, were rigorously established from first principles. Starting with a handful of simple statements, and proceeding inexorably one small step at time, Euclid obtains result after result of genuine depth. Achievements wrung over the ages with great difficulty were made to seem inevitable.

The appearance of the *Elements* in a culture that valued geometry, that puzzled over the different results, and that had worked out the rules of formal reasoning contributed to the explosively creative early years of Alexandria. Extraordinarily talented mathematicians, the most brilliant of whom were Archimedes (287–212 BCE) and Apollonius (262–190 BCE), created far-reaching mathematical results and theories that built on the foundations laid by the *Elements*. In the next two hundred years, mathematics and science would advance much further. Theodosius (160–90 BCE) and Menelaus (c. 70–140 CE) investigated geometry on the sphere. Hipparchus (190–120 BCE) and Erastosthenes (276–194 BCE) brought mathematical geography and astronomy to a higher level. Much of this later work has been lost, or has come to us in a form that makes it difficult to determine what the author was saying.

RIGOR AND THE *ELEMENTS*

Any work as self-consciously rigorous as the *Elements* invites the question of just how rigorous it is. Contrary to the insistence of generations of well-meaning teachers, Euclid does not argue only from axioms and definitions. He tacitly uses other properties.

Consider, for example, proposition 1. It states that given any line segment, we can construct an equilateral triangle that has that segment as a side. The argument goes as follows. Let A and B denote the endpoints of the given segment. By postulate 3, we can draw the circle centered at point A having the segment AB as radius. By postulate 3 again, draw the circle at point B with the same radius.

These circles (see figure 23) meet at two points. Choose one. Call it C. By postulate 1, we can draw the lines AC and BC (see figure 24). Then the triangle with sides AB, BC, AC has to be equilateral because the lengths of AB, BC, and AC are equal (because all are radii of circles with equal radius).

Nice. But what postulate or property says that the circles centered at points A and B must meet?

This does not follow from the postulates or the definitions, and is an obvious gap, noticed almost from the very beginning and mentioned in many commentaries. If the game is to be completely explicit about all assumptions, and to take nothing for granted, what allows Euclid to implicitly assume that two lines or circles that cross each other must have a common point? We need some sort of betweenness principle stating that if a line or circle contains points on different sides of another line or circle, then the two must have at least one point in common.[34] There are other gaps as well. Many proofs, not just the one of the first proposition, leave much to be desired. Moreover, some

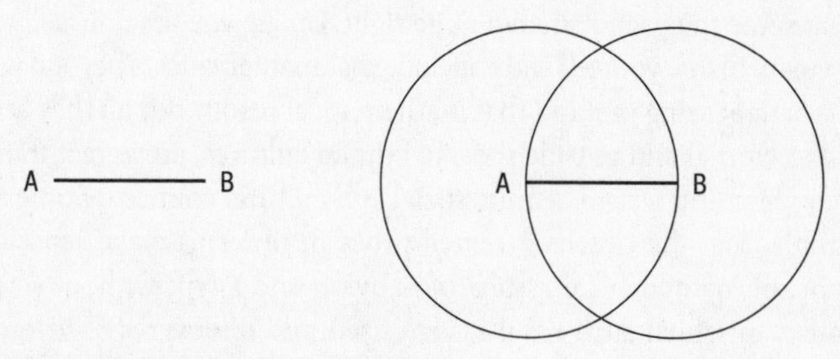

FIGURE 23. *Two circles with the same radius*

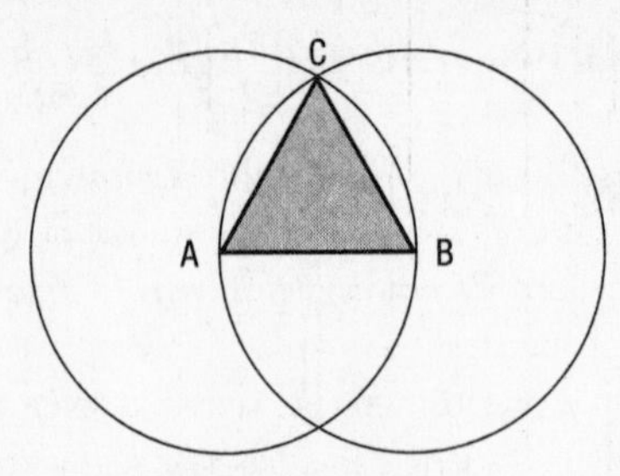

FIGURE 24. Construction of an equilateral triangle

of the postulates are unclear. Does postulate 2 mean that we can extend any line segment forever? Does it mean that we can cut up any segment? And if it means the first, who is to say that the resulting line is unique? And how seriously should we take the definitions? Are they just meant to provide guidance about a word that is essentially undefined (today's, and probably Euclid's, interpretation) or are they supposed to completely specify the object named? In the latter case, just what does the phrase "a breadthless length" mean?

Mathematicians and scholars know that there are gaps in Euclid, and there has been a great deal of discussion over the ages about alternate axioms, or possible additional ones. That has not stopped generations of worshipful schoolmasters, besotted with the majestic order, the accessibility and the patent usefulness of the *Elements* from rushing in and trumpeting it as the finest in human thought. However, to a thoughtful student, the *Elements* can seem less rational than capricious. The insistence that the *Elements* is flawless, and the apex of rigorous thought, turns some students away from mathematics. One wonders how much fear of mathematics stems from the disjuncture between the assertion that Euclid is perfect and some students' intuitive, but difficult to articulate, sense that some things in it are not quite right. Unless you are unusually rebellious, it is easy blame yourself and conclude that mathematics is beyond you.

It is worth bearing in mind that mathematical results, for all they are represented as eternal and outside specific human cultures, are in fact transmitted and understood within definite social and cultural contexts. Some argue, for example, that the Greeks invented proof in order to make sense of the statements of mathematical results of Babylon and Egypt without access to the context in which such results were used and discovered.[35] In order to make use of the results, the Greeks needed to sort out different, seemingly contradictory, computations and re-create them themselves in their own terms. This is certainly plausible. Even within the same civilization, each generation of mathematicians reinterprets and reframes the mathematics of previous generations. To learn mathematics is to reinvent it.

But the ambiguities run deeper still. Alexandria of twenty-three hundred years ago was a culture very foreign to our own. Although it was very advanced mathematically and technologically in its first centuries, large chunks of this learning were lost, and we know almost nothing of the context in which the *Elements* was created. In a provocative book that has just received a marvelous translation from Italian to English, Lucio Russo argues that science, in the modern sense of the term, flourished in Alexandria from 300–150 BCE and was subsequently lost.[36] According to Russo, the geometric parts of the *Elements* were a theory of computation. In his account, the Greeks did computations by first translating them into geometry, then drawing the relevant geometric construction using a ruler and compass, and measuring. Like the slide rule centuries later, the ruler and compass were analogue means of computation, and the *Elements* was a manual of sorts, showing how and for what one could use them.

Russo's argument is quite a way out of the mainstream, and will surely draw some heated criticism. But the obviousness of the gap in the proof of proposition 1 supports Russo's contention that Euclid was creating a mathematical model of what a person could do with a ruler and compass on a piece of parchment, and not aiming for absolute purity. The first axiom states that we can draw a line between any two points, and the second that we can extend any line indefinitely. Taken together, these amount to saying that we have a ruler, but that we are going to ignore any complications that arise from its being too short. The third axiom, positing that we can construct a circle centered at any point of any desired size, says that we have an "ideal compass"—we can assume that it is as big, or small, as necessary. One can imagine Euclid saying that we are going to explore what we can do with a ruler and compass yet without worrying about their physical limitations. It would never occur to him to posit that lines and circles drawn with rulers and compasses intersect in a point if they cross one another. This would seem completely obvious if he had ideal physical objects in mind.

The plausibility of Russo's argument should teach us humility. We don't know the purpose or the intended readership of the *Elements*. But the widespread notion that it was a textbook for schoolchildren may not be accurate. Even the most casual reading of the *Elements* suggests that it was written for grown-ups, not children. The assumption that it was a textbook for pupils associated with the Alexandria Library is just that: an assumption.

THE LONGEVITY OF THE *ELEMENTS*

Its flaws notwithstanding, one cannot read the *Elements* without coming away from it with a genuine admiration for the artfulness with which it is laid out, and the cleverness of the proofs. The relentless advance from the simplest notions to subtle, deep, and beautiful propositions testifies to the efficacy of human reason. Looking back from the vantage of today, it is easy to take the longevity of *Elements* for granted and to concoct after-the-fact rationalizations that explain its survival. Anything this good must endure, one would think. We tell ourselves that the really good books from ancient times are the ones that survived. According to this chestnut, the more authoritative a manuscript was, the more likely it was that it would be copied and recopied. The more likely, too, that it would have been translated into Arabic and survive that way even if the Greek manuscript were lost. Such wishful thinking is reassuring, but far too many books were lost to indulge in it. Euclid was the most famous geometer of ancient times, yet half of his books have disappeared.[37]

The blithe assumption that anything as good as the *Elements* must endure detracts from the miracle of its survival. The *Elements* endured even as the creative energies and quality of Alexandrine scholarship slowly leached away (and they were largely past by the time Julius Caesar torched Alexandria harbor in 47 BCE). It survived the end of Alexandria as a place of learning, an extinction marked by the brutal murder of the Neoplatonist mathematician Hypatia (370–415) in March 415. Threatened by her charisma and the power of her lectures, a frenzied Christian mob stripped her naked and flayed her to death seeking to erase her authority, her beauty, and her learning. They did not succeed. The edition of the *Elements* on which she had worked with her father, Theon (335–c. 405), became the standard after her death. It was the basis of what Caliph al-Mansur (712–775, reigned 754–774) obtained from the Byzantine emperor and which the great Arab translator al-Hajjaj (c. 786–833) would subsequently translate not once, but twice.[38] Arab scholars were as fascinated by the *Elements* as had been the Greeks and the books would be retranslated, recopied, and reedited hundreds of times in the great centers of Arab learning. Countless commentaries, summaries, expositions, and translations were made.

Several hundred years later, the *Elements* and a number of works of Aristotle were among the very first Greek works to be translated back from Arabic into Latin. Gherard of Cremona (1114–1187) seems to have made the first such translation. Johannes Campanus, a chaplain to Urban IV who was pope 1261–1281, retranslated it. The onslaught of translations of classical texts by Gherard and others coincided with the near-spontaneous emergence of our oldest universities in Bologna, Paris, Oxford, Cambridge, Salamanca, and elsewhere in Europe. In a very real sense, Euclid is at the heart of our universities.[39]

The *Elements* was the first scientific book to be published after printing presses emerged in the mid-fifteenth century.[40] Different editions of it were best-sellers in Renaissance Europe. It is widely conceded to be the second-most-read book in human history.[41] However, when you consider that it was translated into Chinese in 1607 and had penetrated the Indian subcontinent by the tenth century, it may well in fact be the most-read book of all times. This is all the more striking when one realizes that its only rivals are the Bible and the Quran.

The popularity of the *Elements* over time suggests that it responds to a deep human need. Many have hailed the salutary effect that studying it has on the development of one's reasoning ability. While in Congress, Abraham Lincoln took it to bed with him every night. Thomas Jefferson advised an inquiring young man that he would find the most useful results in the later books. Others theorized that Euclid was especially useful for young women: As early as 1838, Mount Holyoke College, the oldest women's college in the United States, required its students to own and study either Simson's or Playfair's Euclid.[42]

Despite the survival and the popularity of the *Elements*, we really have no idea what the original looked like. This would be the case for any book that age: Before the invention of printing, documents were copied from one hand to another, with all the errors that are introduced in transcribing texts compounding with each transcription. It is especially so for a book as widely copied as the *Elements*. The complexity in the transmittal of the *Elements* to present day far exceeds that of any other ancient text.[43] Thousands of scholars and teachers have studied it, annotating it carefully and recopying it in ways that strike them as clearer. Different editions, annotations, and translations abound. The original Arabic translations of the Greek have been lost, as have the Greek manuscripts from which they were translated.

For a long time, it had been thought that the standard Arabic edition was actually older than extant Greek editions. However, in 1808, François Peyrard argued that a Greek manuscript copy of the *Elements* in the Vatican Library that Napoleon had pilfered and taken to Paris was actually older than the Arabic edition. The vital clue was a comment that Theon made in his commentary to Ptolemy's *Almagest*, indicating that he had added material to the final proposition of book 6 of the *Elements*. The Vatican manuscript did not contain this addendum. Peyrard went on to correct the then-authoritative Greek edition [that had been prepared by Simon Grynaeus (Basel, 1533)].

In 1883–84, the Danish scholar J. L. Heiberg published a very learned reconstruction, starting from scratch with the original Greek text based on the Vatican manuscript and other manuscripts. There is no question that Heiberg's result is "purer" than the version attributed to Theon and Hypatia, and his is the version most scholars start with today. It is the basis of the standard English translation, prepared by Heath in 1908. In spite of Heiberg's massive scholarship, we do not know how much older the Vatican manuscript is than the version of the *Elements* of Theon and Hypatia. The latter appeared seven hundred years after Euclid's *Elements*. The Vatican version could have appeared several hundred years earlier than the Theonine version and still have absorbed several hundred years' worth of changes. For all we know, Heiberg's reconstruction may differ greatly from Euclid's original.

What is most inspiring about the *Elements* is not its majesty, but its contingency. The text itself is not what matters and indeed, appearances to the contrary, is not what has endured. What matters is the curiosity that things geometric engender, the willingness to question received wisdom, and the way in which human knowledge builds on work of others. Virtually every word and every line of the *Elements* has received extended attention from thousands of commentators. Alternate phrasings have been proposed, different proofs mooted. Theorems are known in different places of the world by different pet names. Proposition 5 is known in England as the *pons arsinorum*, "Asses Bridge." Proposition 47, known nowadays as the Pythagorean theorem, was in times past referred to as the "Wedding Theorem" or the "Theorem of the Bride."[44] The most complete commentary to have survived from ancient times is Proclus's (410–485) from Rome. He, in turn, seems to have had at hand no fewer than four earlier major commentaries, all lost excepting fragments, underscoring again the fragility of individual works.[45] The *Elements* is a leitmotif for the human enterprise, for our dependence on one another, and for the unreliability of individual achievement. With regard to the Poincaré conjecture, what matters most about the *Elements* is the fifth postulate and how it has given way to our current understanding of space.

6

The Non-Euclideans

Understanding what it means for space to be curved is critical to Perelman's proof of the Poincaré conjecture. Understanding curvature, in turn, requires complete clarity about what it means to be straight and to be flat. For this we need not just Euclid's *Elements*, but the sophisticated understanding of it that has evolved over two millennia. More precisely, the seemingly innocuous notions of straightness and flatness require unraveling the riddle implicit in the fifth postulate of the first book of Euclid's great work.

The Fifth Postulate

From the very beginning, Euclid's fifth postulate has raised hackles:

> If a straight line falling on two straight lines makes the interior angles on the same side less than two right angles, the two straight lines, if produced indefinitely, meet on that side on which the angles are less than the two right angles.

Look at the list of five postulates reproduced on page 49. The first four take less than a line to state. Not so the fifth. Euclid must have thought long and hard before adding it to his list, and he cannot have been satisfied by it. He puts off using it as long as possible (until proposition 29), and for good reason. From the outset, it was attacked as being too complicated. The other postulates are clean, easy to state, and manifestly self-evident. The fifth, by contrast, seems complicated. Ugly. You have to read it a couple of times to figure out what it is saying.

To rephrase it in more familiar terms, recall that a complete rotation consists, by convention, of 360 degrees. A quarter turn, or right angle, is 90 degrees, and two right angles make a turn of 180 degrees. The fifth postulate states that if we take a fixed line, which we may as well assume to be vertical, and then take an additional two lines that cross it in such a way that the sum of the interior angles (A and B in figure 25) on the right side is less than 180 degrees, then the two lines eventually have to intersect in a point on the right side.

"This is no postulate, it is a theorem," complained Proclus, and he tried to prove it. Frustrated, he recounted the great geographer Ptolemy's efforts at a proof. Ptolemy devoted a book, now lost, to trying to prove the fifth postulate. As Proclus pointed out, and as Euclid undoubtedly knew, the fifth postulate is equivalent to the statement that we usually call the "Parallel Postulate"[46]:

> Given any straight line l and a point P not on it, there is exactly one line parallel to l that passes through P.

This axiom is often called Playfair's postulate, since he took a variant as his fifth axiom in his edition of Euclid.

Euclid did have the benefit of some serious discussion among the Greeks about the existence of parallels. Most of this discussion has been lost, but we know of it because Aristotle complained about the circularity of many of the

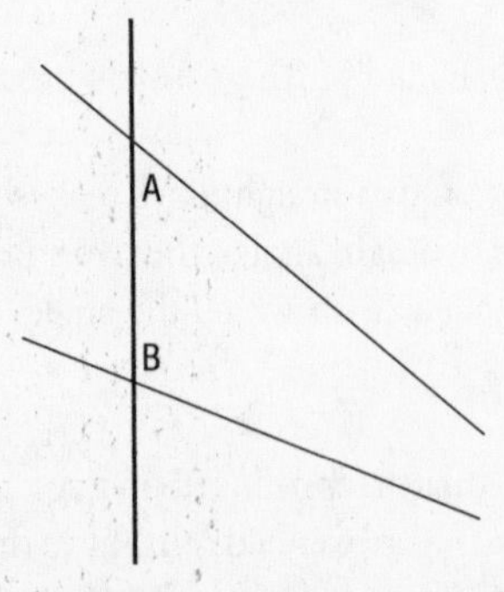

FIGURE 25. *Since the sum of angles A and B is less than 180 degrees, the two lines crossing the vertical line must eventually intersect at a point to its right.*

arguments involving parallels. He pointed out that many arguments showing the existence of parallels use facts that are actually equivalent to the existence of parallels, and hence tacitly assume what they are trying to prove. In his words, "If A follows from B and B follows from C and C follows from A, then you can't say that you have demonstrated A because C is true."

Nonetheless, one has to admire Euclid's courage in listing the fifth postulate. Most people would have got hung up on it. Instead, Euclid bit the bullet, got on with it, and left us a masterpiece.

The Arabs, too, became obsessed with the fifth postulate, and tried to deduce it from the other postulates, or to replace it with something else. All to no avail. They did, however, introduce many new mathematical techniques that simplified computation and made algebra independent of geometry.

Nonetheless, questions about the fifth postulate lurked. Could one deduce it from the other postulates? Or was there, perhaps, a more evident statement that might be taken as an axiom? Science as we know it emerged in the seventeenth century, and its development was accompanied by an explosion of mathematical discovery. In this febrile atmosphere, the fifth postulate became an obsession for many. John Wallis, Newton's predecessor, surveyed the literature on it in his book *De postulato quinto*. From 1607 to 1880, well over a *thousand* books and memoirs were published devoted to the fifth postulate. Very serious attempts were made to prove it, and different proofs announced, but none bore up to careful scrutiny.

Many attempts at a proof were magnificent. Among the most spectacular, Gerolama Saccheri (1667–1733), a Jesuit priest and professor at the University of Pavia, investigated a number of postulates equivalent to the fifth, and assumed they were false. For instance, he carefully showed that the fifth postulate was equivalent to the sum of the angles of a triangle being equal to the sum of two right angles (that is, 180 degrees), and he considered the two cases where the sum of the angles was greater than two right angles, and where it was less, hoping to obtain a contradiction. He even convinced himself that he had discovered one.[47]

With the Enlightenment and the rise of science, new universities were founded. One of the greatest, the University of Göttingen, was founded in 1737 by George II, the king of England and elector of Hanover.[48] Created in the spirit of the Enlightenment, it was among the first institutions in which the science faculties were on an equal, as opposed to subordinate, footing to the faculty of theology. One of the professors at Göttingen, mathematician Abraham

Kästner (1719–1800), who was interested in the history and foundations of mathematics, prefaced his widely read four-volume book *Mathematische Anfangsgründe* as follows:

> The difficulty that arises in the theory of parallel lines has occupied me for years. I used to believe that it had been entirely removed by Hausen's *Elementa matheseos* [1734]. The former preacher to the French congregation in Leipzig, Mr. Coste, shook my complacency when, during one of the walks that he granted me, he mentioned that, in the above-mentioned work by Hausen, an inference is made that does not follow. I soon discovered the mistake myself, and from that moment exerted myself either to remove the difficulty or to find an author who had removed it; but both efforts were in vain, although I soon assembled a small library of individual writings of works on the first principles of geometry where this topic was considered especially closely.[49]

Instead of the fifth postulate, Kästner adopted an equivalent axiom of Wallis's.[50] Kästner's student, G. S. Klügel, wrote a thesis in 1763 that examined some thirty proofs of the fifth postulate and found them all defective. He concluded: "To be sure, it may be possible that non-intersecting lines diverge from one another. We know such a thing is absurd, not by rigorous inference or clear concepts of straight and curved lines, but rather through experience and the judgment of our eyes."[51]

Klügel's dissertation interested Kästner's friend, the gifted Johann Heinrich Lambert (1728–77), in the subject. Lambert was born to a large family, and necessity forced him to follow his father into his business as a tailor. But he continued to study on the side, eventually securing a position as a tutor in the family of a Swiss nobleman, which gave him time for research. Lambert was as eccentric as he was brilliant. Oddly dressed, he decided that the best way to talk to someone was not standing face to face, but at right angles to one another. He was enthusiastically nominated to a position in the Prussian Academy of Sciences in Berlin by the legendary Swiss mathematician Leonhard Euler (1707–83). But Frederick II balked, reportedly telling a friend after meeting Lambert that he had just met the greatest blockhead in all of Prussia. The monarch soon changed his mind, learning to value Lambert's insight. The colorful Lambert would go on to make major contributions to optics, cosmology, philosophy, and mathematics.[52] Not least among these was a book *Theorie der Parallellinien* on the parallel postulate that he wrote in 1766. He did not publish it because he couldn't settle the issue to his own satisfaction. However, it was published posthumously in 1788.

Like Saccheri, the Jesuit of Pavia, Lambert explored the consequences of assuming that results equivalent to the fifth postulate were false. In particular, he obtained formulas for the area of a triangle in terms of the sum of their angles in the cases that the sum of angles of a triangle is greater than two right angles, and less than two right angles. He noted that on the sphere, if one takes straight lines to be great circles, then triangles have more than 180 degrees and his formula correctly gave the area of triangles. (This result, including the formula, was actually known to the Greeks.) He wondered if the sum of the angles could be less than two right angles on an appropriate imaginary surface (and even mentioned spheres whose radius involves square roots of a negative number).

THE END OF THE EIGHTEENTH CENTURY: *FIN-DE-SIÈCLE 18ÈME*

By 1800, Enlightenment ideas were everywhere. The seeds sown by Descartes, Galileo, and Newton had come to fruition. The air rang with new intellectual confidence and a faith in human reason and in science. For the first time since the Ionian philosophers and the Alexandrian school, the universe was seen to be comprehensible. The scientific societies that had formed earlier, many with the support of enlightened despots, had become well established. In addition to the societies and the older universities, new institutions of higher learning had been founded to supply the needs of an increasingly complex society.

The scientific ideas of the times and their reflections in philosophy and the arts shaped a worldview in which the universe ran in accordance with a series of mathematical laws knowable to man. "*Sapere aude!* Have courage to use your own reason! . . . That is the motto of the Enlightenment," wrote the influential German philosopher Immanuel Kant (1724–1804).[53] "We hold these truths to be self-evident" proclaimed the 1776 American Declaration of Independence. Some truths are self-evident, and all are accessible to reason and deduction. Among such truths, according to Kant, are the propositions of Euclid's geometry, and the self-evident truth that these propositions apply to our universe. On stages throughout Europe, the hero and heroine of Mozart's opera *The Magic Flute* sought admission to a brotherhood, strongly reminiscent of the Pythagoreans, preaching universal love and reason.

Against such a background, the uncertainty surrounding the fifth postulate was like a low-grade head cold, persistent and even somewhat embarrassing. No one seriously doubted the postulate's truth, although some allowed that it might be necessary to invoke some simpler, yet-to-be-discovered principle. "*Un scandale*" huffed the always excitable French mathematician and encyclopaedist, Jean d'Alembert (1717–83). While everyone expected that the uncertainty would be settled sooner or later, most realized that investing time in seriously investigating the issue promised a slim payoff.

D'Alembert's compatriot, Adrien–Marie Legendre (1752–1833), was one of the really able mathematicians to attempt to take on the fifth, so to speak. His text *Eléments de géométrie*, simplifying and modernizing Euclid's *Elements*, appeared in 1794. It became the leading advanced treatment of elementary geometry for one hundred years, and would go through edition after edition. Over the next thirty years, Legendre would try to establish the parallel postulate, different attempts appearing in different editions. He failed but died firmly convinced of its truth.

As the eighteenth century drew to a close, Enlightenment thought, rapid economic change, and an ever-more-educated populace increasingly challenged received ideas. The Declaration of Independence espoused the rights of every individual, and the right to freedom and education. The beginnings of the social changes wrought by the Enlightenment were giving way to the more ungoverned, explosive changes that would mark our times. Social turmoil swept Europe with different unpredictable effects. New possibilities opened up, but the old certainties were gone. Horrified Europeans watched as the French Revolution (from 1789 on) took hold and washed away the old order in a bath of blood. Nothing was sacred. The armies marshaled by relatives of the executed French royals, Louis XVI and his wife Marie Antoinette, attacked France, thereby creating the conditions that allowed Napoleon to counterattack, to ascend to power, and to ultimately conquer most of Europe (from 1805 on).

GAUSS, LOBACHEVSKY, BOLYAI

Three individuals, Johann Carl Friedrich Gauss (1777–1855), Nikolai Ivanovich Lobachevsky (1792–1856), and János Bolyai (1802–60), would finally clarify the role of the fifth postulate and the riches it concealed.

Gauss was the most famous mathematician of the early nineteenth century and one of the greatest mathematicians of all time.[54] His father was a laborer with an elementary school education, who could not break out of a series of menial jobs. His mother was a maid with even less education. The father's family had moved from the farm to the city of Brunswick in what is now Germany. A few years earlier there would have been no chance of someone like Gauss receiving any substantial education. Even with fifty students in his elementary school class, he stood out. He had already taught himself, seemingly without parental intervention, how to read and write before entering school. It was his (and our) good fortune that the teacher's assistant, Martin Bartels, though only eight years older than Gauss, had studied mathematics at Göttingen. Bartels paid especial attention to Gauss, and together he and the teacher arranged for the boy to attend the gymnasium, one of the rigorous German high schools geared to able students who planned to pursue advanced studies. Three years later, Gauss was introduced to his prince, the Duke of Brunswick-Wolfenbüttel, who subsequently granted Gauss a regular subsidy. Such subsidies to promising, but financially needy, young people were not uncommon, and were the forerunners of today's academic scholarships. They were a prudent investment given the government's need for an ever-more-educated workforce. The stipend allowed Gauss to continue his studies, first at a newly established, elite science-oriented academy[55] in Brunswick (1792–95), and subsequently at the University of Göttingen (1795–98), about sixty-five miles south of Brunswick, in the then-foreign state of Hanover.

While at Göttingen, Gauss met Farkas Bolyai (1775–1856), the future father of János Bolyai. They were an unlikely pair. Farkas (also called Wolfgang) came from a once-wealthy, downwardly mobile family with a long history of fighting the Turks. Farkas's father had held onto a small estate, but the family money was gone. Farkas had left school at age twelve and owed his attendance at university to having been appointed as tutor to eight-year-old Simon Kemený, son of Baron Kemený. Farkas and Simon became close friends, and the baron sent Farkas with Simon to study at Göttingen. Gauss and Farkas both enrolled the course of Abraham Kästner, the same mathematician who had been so interested in the fifth postulate. Kästner would have been in his late seventies at this time, with his most dynamic years of lecturing behind him. Gauss skipped most of Kästner's lectures, finding them too elementary, and reportedly enjoyed making fun of Kästner. After class, Farkas and Gauss discussed Euclid's axioms and the possible independence of the parallel postulate, as well as other mathematics. Farkas was hooked, and Gauss retained a lifelong interest in the issue. Doubtless, Gauss had heard previously of the questions involved in the fifth postulate from Martin Bartels, who had also taken Kästner's classes.

Gauss and Bolyai completed their studies in 1798 and returned home to uncertainty. At the duke's behest, Gauss submitted a doctoral dissertation to the University of Helmstedt, the local university the duke supported, and received his degree in 1799. His first book, *Disquisitiones Arithmeticae,* appeared in 1801 and was instantly recognized as a masterpiece. But it was not until the following year that he became really famous. In June 1801, one of the leading German astronomers had published the orbital positions of the asteroid, Ceres. It had just been discovered on January 1, and had disappeared on February 11 behind the Sun. It was widely expected to reappear later in 1801 or early in 1802. But where? At the time, the central problem of computational observational astronomy was the computation of orbits of heavenly bodies from a small number of astronomical observations, each containing errors. Gauss used a technique that he had invented, but not published, to compute a predicted orbit that differed greatly from those of other astronomers. When Ceres reappeared where and when he predicted, fame followed.

Gauss immediately received an offer to become director of the observatory at St. Petersburg, but Brunswick had already started to build an observatory for him. The German astronomers worked hard to keep Gauss, and he eventually received an offer of a position from the University of Göttingen. They promised to build an observatory for him there, too, and to supply a talented assistant. Gauss accepted in 1805, married, and turned down offers from St. Petersburg and Bavaria. His timing could not have been better. At age seventy, his patron, the Duke of Brunswick-Wolfenbüttel, had been put in charge of the Prussian forces, and was mortally wounded in the battle of Jena against Napoleon. Although Hanover changed hands, becoming part of the French-dominated Kingdom of Westphalia, the French were able administrators, and the University of Göttingen continued to thrive.

Things went very differently for Gauss's old friend Farkas.[56] Farkas's patron, the baron, suffered some financial reverses and only sent money for his son Simon to return home. Broke, Farkas stayed on in Göttingen for a year, living on borrowed money and the charity of friends. Finally, a friend sent him enough money to pay off his debts and he then walked all the way back to Hungary. There, he reluctantly took an ill-paying, time-consuming but secure

FIGURE 26. Gauss's observatory today

job teaching mathematics, physics, and chemistry at the Calvinist college in Marosvárához. He married in 1801 and his son, János, was born a year later, in December 1802. His wife appears to have suffered from extreme anxiety and became increasingly difficult to live with as her health deteriorated over the years. Farkas took on extra duties to supplement his earnings: he wrote and published dramas, ran the college pub, and designed tiles and cast-iron stoves. Throughout it all, he continued to work on mathematics in his spare time.

Farkas poured his efforts into his son, who was brilliant, and whom Farkas wanted to become a mathematician. He homeschooled János until the age of nine, having students from the college tutor him, and continued to supplement his mathematics instruction after Janos had entered the gymnasium (college-preparatory secondary school). At age thirteen, János could play the violin like a professional, had mastered calculus and analytical mechanics, and could speak several languages. But Farkas did not have the money to secure a first-class university education for János. He wrote to Gauss in 1816, suggesting that János go live with Gauss and study mathematics. Nothing came of this, and Farkas and János decided that the best they could do would be to have János attend the Royal Engineering College in Vienna. János finished the seven-year course of studies in military engineering in four and subsequently served eleven years in the Austro-Hungarian army, where he was reputed to be the best swordsman and dancer in the entire Imperial Army.[57]

The same winds of Enlightenment change that were lifting Gauss's fortunes and depressing the Bolyai family's were blowing just as hard in the easternmost reaches of Europe. Nicolai Ivanovich Lobachevsky's father, a clerk in a land-surveying office in Nizhny Novgorod in Russia, died when Lobachevsky was only seven, leaving his wife to fend alone with three young children and no money. She moved the family to Kazan, at the edge of Siberia, where the children attended gymnasium on state scholarships that financed the education of poor but talented boys. Lobachevsky went on to study at Kazan State University, which had just been founded in 1805 as a result of one of the reforms of Czar Alexander I of Russia. Among the first professors appointed at the university was the same Martin Bartels who had been the assistant in Gauss's elementary school class. A gifted teacher, Bartels lured Lobachevsky, who was originally intending to go into medicine, into mathematics. Fatefully for the future, Bartels's course on the history of mathematics used a text that discussed the fifth postulate.

Lobachevsky graduated in 1811, and ascended through the ranks to become a full professor eleven years later. Service on a committee led to his becoming dean of the mathematics and physics department, then head librarian, then head of the observatory, and ultimately rector of the university. A talented administrator, he guided the university through the difficult times in the czar's later years (1819–26), which saw a return to despotism, a distrust of Enlightenment thinking, and the departure of some of the best professors, among them Bartels. In the period of increased tolerance that accompanied the accession of Nicholas I in 1826, Lobachevsky became the leading innovator at the university, restoring academic standards and faculty morale.

By the 1820s, Lobachevsky, in Kazan, and Farkas Bolyai, in Marosvásárhely, had jobs remote from any major mathematical center, and were spending much of their free time investigating the fifth postulate. Gauss, too, maintained a strong interest in geometry. In a letter to Bolyai, he pointed out that a proof that Bolyai proposed for the fifth postulate was incorrect. He wrote a book review in 1816 criticizing several alleged proofs that the fifth postulate can be deduced from the others.

Gauss also had a strong practical bent. After 1815, all major states in central Europe were funding geodetic surveys, precise surveys that mapped large areas making corrections for the curvature of the Earth. In 1818, Gauss became the director of a large project to accurately survey Hanover and Bremen and to connect these surveys with a Danish one. The flat and heavily forested coastal lands created many difficulties. One could not get enough high points from which to carry out triangulations and measurements with any accuracy. The completion of this survey occupied much of Gauss's time from 1818 to 1832. In the process, he made major contributions to geodesy.

The work also stimulated Gauss's interest in differential geometry, the subject that uses calculus to study geometry. He published a prize-winning paper in 1823, discussing one-to-one correspondences between surfaces that preserve angles. Five years later, his exquisite little book *Disquisitiones generales circa superficies curvas* laid out his geometric ideas, substantially advancing the work of Euler and others. The book made no mention of the fifth postulate and was solid mainstream mathematics. In it, Gauss worked out the equations for measuring distances on arbitrary surfaces in three space. He defined a notion, now called the *Gaussian curvature*, that measures how a surface curves, and showed that this quantity can be calculated from measurements taken only on the surface—one does not need to be able to see off the surface. He also discovered that there is a relationship between the areas of triangles on a surface and the average curvature inside the triangle.

BEYOND THE FIFTH

By the mid-1820s, and possibly earlier, Gauss seems to have convinced himself that one can have geometries in which the fifth postulate does not hold. In November 1824, he wrote to his friend Taurinus that

> The assumption that the sum of the three angles of a triangle is less than 180° leads to a curious geometry, quite different from ours [i.e., Euclidean geometry] but thoroughly consistent, which I have developed to my entire satisfaction, so that I can solve every problem in it excepting the determination of a constant, which cannot be fixed a priori. . . . the three angles of a triangle become as small as one wishes, if only the sides are taken large enough, yet the area of the triangle can never exceed, or even attain a certain limit, regardless of how great the sides are.[58]

There was no way that Gauss was going to publish anything like this. Although world famous and well-connected mathematically, he maintained a sort of self-imposed isolation. Conservative, aware of his own good fortune, and always feeling a little tenuous in his position, he took particular care not to offend those who had power over him. Gauss's relationship with his own father was poor. Gauss's first wife had died in 1808 only three years into their marriage, while giving birth to their third child. He married her best friend, by whom he had another three children, but that marriage was never as warm as the first. His relationship with his sons was tangled. He seemed aloof, was somewhat of a loner, and had few friends. Even the celebrated German explorer and statesman Alexander von Humboldt, who had warm relationships with many mathematicians, characterized Gauss as glacial.

Gauss knew that publishing his results on the fifth postulate would have caused a sensation, and he did not want the publicity or the bother. Over the years, interest in the fifth postulate made its way into philosophy and the popular press. He was aware of how many books and papers had been devoted

FIGURE 27. *Carl Friedrich Gauss*

to the subject, and how many cranks lurked waiting to surface. He also had a healthy suspicion of philosophers: "When a philosopher says something that is true, then it is trivial. When he says something that is not trivial, then it is false."[59]

Meantime, János Bolyai told his father in 1820 that he was working on the fifth postulate. Alarmed, his father replied trying to dissuade János:

> I implore you to make no attempt to master the theory of parallels; you will spend all your time on it.... Do not try... either by the means you mentioned or any other means.... I passed all through the cheerless blackness of this night and buried in it every ray of light, every joy in life. For God's sake, I beseech you, give it up. Fear it no less than sensual passions, because it too may take all your time, deprive you of your health, peace of mind and happiness in life.[60]

Strong words, reflecting the elder Bolyai's bitter experience.

At first, János tried to replace Euclid's postulate with another statement that could be deduced from the others; he did in fact give up on this approach within a year, but otherwise ignored his father's entreaties. He began to work seriously on what happens if one assumes that the fifth postulate did not hold. His notebooks show that he had begun to develop what we now know as hyperbolic geometry. In 1823, he wrote to his father that he was in the process of creating "a new, another world out of nothing..." He seems to have finished by 1824.

After some initial skepticism, the elder Bolyai became convinced of the value of his son's work. He urged János to write it up as an appendix to Farkas's magnum opus, the *Tentamen*, a rigorous and systematic foundation of geometry, analysis, arithmetic, and algebra published in 1831.[61] Farkas sent a copy to Gauss, who on reading the Appendix wrote to his friend Gerling, "I regard this young geometer Bolyai as a genius of the first order." To Bolyai senior, he wrote:

> Now a few words about your son. If I begin by saying that I cannot praise this work, you will be astonished at first, but I cannot do otherwise. For to praise it would be to praise myself. The entire content of the essay, the path that your son follows, and the results he obtains, coincide with my own discoveries, some of which date back 30 to 35 years. I am stunned. Concerning

> my own work, of which I have not to date put much to paper, my intention was not to publish during my lifetime. Most people do not have clear ideas about the questions of which we are speaking, and I have found very few people who would be especially interested in what I have to say on the subject. To take such an interest, one first must have thought carefully about the real nature of what is wanted and upon this almost all are quite uncertain. On the other hand, I had intended to write all this down later so that at least it would not perish with me. It is therefore a pleasant surprise for me that I am spared this trouble, and I am especially glad that it is just the son of my old friend who takes precedence to me in this matter.[62]

Farkas was happy with the reply, but János was shattered. His mental health began to deteriorate. He became irritable and unstable, and was pensioned off in 1833. As if that were not bad enough, it turned out that Bolyai was not even the first to publish. Lobachevsky had also begun working on the fifth postulate in the early 1820s, and realized that the geometry you get when you deny the fifth postulate seems to make perfect sense.

Lobachevsky's ideas were rooted in his opposition to Kant's transcendental idealism, which maintains that such ideas as space, time, and extension are known a priori, and that the mind imposes order on sense experience. For Lobachevsky, space was an a posteriori concept, derived by the human mind from external experience. He announced his results in 1826 and published his theory of non-Euclidean geometry in 1829. He originally submitted the paper for publication in the St. Petersburg Academy of Sciences, which had a relatively wide readership. However, the leading Russian mathematician of that day, Mikhail Vasilevich Ostrogradsky, who was well known in western Europe, rejected the paper. So Lobachevsky published, in Russian, in a local general periodical and, from 1835 to 1839, still in Russian, in the Kazan academic transactions. Not surprisingly, this work went largely unnoticed.

So what happened? In a word, nothing. At least for a while.

Lobachevsky continued in administration, finding time for mathematics on the side. His administrative skills were widely appreciated. He had pulled Kazan State University through a number of crises: reconstructing the faculty after 1826, saving lives during the cholera epidemic in 1830, rebuilding university buildings after a devastating fire in 1842. His efforts in popularizing science and in modernizing primary and secondary education in the region were Herculean. Lobachevsky married a much younger woman from a wealthy family in 1832, and had seven children. Financial difficulties, ill health, increasing blindness, and the death of his eldest son marred Lobachevsky's later years.

Sadly, too, Lobachevsky's mathematical work was not generally recognized during his lifetime. Nonetheless, he continued to publish. He also published an account of his work in French in the leading mathematical journal of the time in 1837, and a book in German in 1840 on the theory of parallels. The latter greatly impressed Gauss, who subsequently arranged for Lobachevsky's election to the Göttingen Academy of Science, and learned Russian to read the rest of Lobachevsky's work. Typically, however, Gauss made no public statement endorsing the work. Lobachevsky died, blind, poor, and brokenhearted on February 24, 1856.

Farkas Bolyai retired from teaching in 1851, and died after a series of strokes on November 20, 1856. His first wife had died in 1821, and he was survived by his second wife. His will poignantly summarizes his life after returning from Germany as follows:

> Until I returned from Germany it had been morning with the prospect of beautiful days which, after some days laden with fire and ice, turned into raining from the permanently overcast sky until this recent snowfall.[63]

In 1833, János Bolyai retired to the family estate inherited from his paternal grandmother. He entered a common-law relationship with a woman of whom his father did not approve, and relations between father and son deteriorated. He continued to pursue mathematics, but far out of the mainstream. Gauss alerted the Bolyai senior to Lobachevsky's work, which is probably how János learned of the 1829 paper in 1846. János worked his way through it, making careful notes and jotting down his tortured thoughts. His admiration for Lobachevsky's arguments was punctuated by dark speculation that Lobachevsky did not exist and that Gauss had engineered the whole charade to deprive him of credit for his work. János married his partner in 1849 after the declaration of Hungarian independence, left her in 1852, and turned from mathematics to creating a general theory of knowledge. He died of pneumonia on January 27, 1860, at the age of fifty-seven. He had never published again after the appendix to his father's book, but left more than 20,000 manuscript pages of mathematical work. These are now in the Bolyai-Teleki Library in the city of Tirgu-Mures.

Even by 1850, there was little recognition that geometries were possible in which the fifth postulate did not hold. Had Gauss spoken out, things would have been different, but Gauss's last years were marked by further withdrawal from the scientific world. He died in Göttingen in 1855 without having published anything on the fifth postulate.

In time, the results of Gauss, Bolyai, and Lobachevsky would enter the mathematical mainstream. The posthumous publication of Gauss's correspondence and scientific notebooks made it clear that Gauss had discovered non-Euclidean geometry first, and hastened the acceptance of Bolyai's and Lobachevsky's work. In a final irony, Gauss's priority led to some unfounded, and now thoroughly discredited, speculation that both Bolyai and Lobachevsky were subtly influenced by Gauss: Bolyai through his father, and Lobachevsky through Bartels. Tom Lehrer, a mathematician and songwriter, who wrote bitingly satirical and very humorous songs during the Vietnam protest era, captured the faint hint of opprobrium in a marvelously funny (albeit unfair) song that he sings in heavily fake-Russian-accented English:

Who made me the genius I am today,
The mathematician that others all quote,
Who's the professor that made me that way?
The greatest that ever got chalk on his coat.

One man deserves the credit,
One man deserves the blame,
And Nicolai Ivanovich Lobachevsky is his name.
Hi!
Nicolai Ivanovich Lobach—

I am never forget the day I first meet the great
Lobachevsky.
In one word he told me secret of success in
mathematics:
Plagiarize!

Plagiarize,
Let no one else's work evade your eyes,
Remember why the good Lord made your eyes,
So don't shade your eyes,

But plagiarize, plagiarize, plagiarize—
Only be sure always to call it please "research."

And ever since I meet this man
My life is not the same,
And Nicolai Ivanovich Lobachevsky is his name.
Hi!
Nicolai Ivanovich Lobach—

EUCLID'S LEGACY

The unsuccessful efforts to prove the fifth postulate in the eighteenth century had resulted in a clear understanding of the many results that were in fact equivalent to it. Among them are the following:

The Euclidean Package:
Statements Equivalent to the Fifth Postulate

1. Given a line l and a point P not on the line, there is precisely one line through P in the plane determined by l and P that does not intersect l.
2. The sum of the angles in a triangle is 180 degrees (that is, two right angles).
3. The ratio of the circumference to the diameter of a circle is the same for all circles, no matter how large or how small.
4. Given any triangle, there are arbitrarily large and small triangles with the same angles and whose sides are in the same proportion to one another.
5. The Pythagorean theorem.

If we accept the first four postulates and one of these results, then the fifth and the other results follow. If we accept Euclid's five postulates, then all of the boxed results follow. The work of Gauss, Bolyai, and Lobachevsky showed that geometries were possible in which the fifth postulate, and hence all of the results in the box, do not hold. This justified Euclid's inclusion of the fifth postulate as an axiom. It cannot be proved from the first four postulates. To get the familiar plane geometry known since Babylonian times, we have to assume (that is, accept without proof) that the fifth postulate, or something equally complicated, is true.

But the true nature of Euclid's postulates, and the fifth in particular, would not be understood for several decades more. Yes, one could not deduce the fifth postulate from the others. And, yes, other geometries in which the fifth postulate did not hold existed. But these other geometries were not just logical curiosities, anomalies springing from the odd failure of the first four axioms to adequately capture reality. These other geometries were every bit as real, and every bit as valuable, as the familiar plane geometry. All this was clear if one looked at things the "right" way, but this right way was much broader, much more capacious, and entirely different from the way that anyone prior to 1850 looked at geometry. The radical change in view that would clarify everything and begin our modern understanding, was laid out in 1854 in the probationary lecture of Gauss's shy, brilliant student Bernhard Riemann. This lecture was one of the greatest moments in the history of science, and is vital to our understanding of Poincaré's work and all of modern geometry and topology.

7

Bernhard Riemann's Probationary Lecture

Those who crowded the MIT auditorium in April 2003 to hear Grigory Perelman speak knew that they had ringside seats in an intellectual revolution. Not so those who gathered in the lecture hall at the University in Göttingen on the tenth of June 1854 for Bernhard Riemann's lecture *On the Foundations that underlie Geometry.* The occasion was Riemann's Habilitation lecture, the formal lecture that is the final test in the long process, inherited from medieval times, by which a candidate qualified to teach at a university. Formulating his thoughts in words did not come easily to Riemann, and he had worked hard on the lecture for the preceding six months. He had taken pains to make it seem comprehensible to all present, but his real audience was one man: Carl Friedrich Gauss, his advisor. Few individuals present grasped its significance or its boldness. But Gauss did.

The speech completely recast three thousand years of geometry, and did so in plain German with almost no mathematical notation. It would not be published until after Riemann's death, barely a decade later, and it would take another decade or two to work its way into the mainstream mathematical consciousness. It would give rise to differential geometry as we know it. Without it there would be no general relativity, and much of Poincaré's work, and all of Perelman's, would have been unthinkable. Unusually agitated afterward, Gauss confided to a colleague that the lecture surpassed all his expectations. He knew that he had just witnessed an unprecedented intellectual tour de force. With just one year left to live, Gauss had been privileged to glimpse the future.

Nowadays, thanks partly to the Clay Institute and to several recent books, Riemann has become better known to the general public.[64] Although his probationary lecture ranks as one of history's greatest examples of a scientific event underappreciated in its own day, Riemann's work has been revered by mathematicians. What makes this especially extraordinary is that Riemann's collected works amount to fewer than six hundred pages, barely the length of a good nineteenth-century novel. Tolstoy's *War and Peace* is over twice as long. Riemann bloomed relatively late and did not live to see his fortieth birthday, but he revolutionized virtually everything he touched.

RIEMANN'S TIMES

Born in 1826, Riemann was one of six children of a penniless Protestant parson. Tuberculosis was in the family, and Riemann was never really well. His mother died when he was seven, and his father raised the children. His brother died before him and three of his sisters died young. Riemann was studious, devout, and shy. Because of his timidity, he was never really comfortable outside his own family and found lecturing an ordeal. How did a poor parson's son, groomed to study theology, find himself lecturing one afternoon to a group that included Gauss, the world's most famous mathematician, on a topic so explosive that Gauss himself would not dare talk about it? As Riemann surely knew, the audience included a philosophy professor enamored of Kant and likely to jump all over any statement calling into question the primacy of Euclid. One can only guess at the source of his courage, and we indulge in a little speculation at the end of the next chapter.

Riemann's training and background were the result of a constellation of social and economic forces that still shape our mathematics and science today, and that are important to understand. He was homeschooled by his father until age fourteen in the village of Quickborn in Hanover, Germany, after which he boarded and attended gymnasiums in nearby towns. He left to study theology at the University of Göttingen in 1846.

Germany of the time was a pressure cooker waiting to blow. Following the defeat of Napoleon, the Congress of Vienna in 1814–15 had successfully returned Europe to the familiar balance of power that had obtained before the tumult of the French Revolution. A loose German Federation of thirty-five sovereign states and four independent free cities was formed with the shared goals of strengthening the position of ruling monarchs, retention of privileges for aristocracy and the Church, and stamping out pesky Enlightenment ideas. Liberal Napoleonic reforms were rolled back, censorship was imposed and a free press suppressed.

But the worm had buried deep into the reactionary apple. The repressive order was unstable, and tension built for thirty years. Most of the population still lived in rural areas, but very few owned the land they farmed. Increasing numbers moved to towns and cities. Sporadic uprisings over wages, living conditions, and lack of social liberty were put down quickly and often brutally, giving rise to further agitation. Even sleepy Göttingen in Hanover was not untouched. Except for the Napoleonic years, when it was part of the ersatz Kingdom of Westphalia, Hanover had been ruled by the British monarch. With the accession of Victoria to the British throne in 1837, all of this changed. Hanoverian law did not recognize the legitimacy of female rulers, so the Duchy of Hanover passed into the hands of the young queen's arch-conservative uncle, the Duke of Cumberland. He promptly revoked the relatively liberal Hanoverian constitution. Seven eminent Göttingen professors who signed a letter of protest were summarily dismissed, creating a furor. It soon spread beyond the bounds of Hanover, for in this period of the nineteenth century, the railroads had linked together much of Germany and, in the process, helped forge a nascent awareness among Germans that they shared a common language and heritage. Throughout Germany, student organizations took the motto "*Ehre, Freiheit, Vaterland*" (honor, freedom, fatherland).

The coupling of civil liberties with nationalist feeling was new and powerful. Food shortages caused famines in 1847 and further unrest. In March 1848, the overthrow in France of the king and the subsequent declaration of the Second Republic triggered revolution in Germany. Massive demonstrations in different German states resulted in the adoption of constitutions ensuring civil liberties and democratically elected governing bodies. Even the two most powerful states, Austria and Prussia, were deeply affected.

This was the backdrop against which Riemann's most formative years occurred. Shortly after arriving in Göttingen in 1846, he took a number of mathematics courses, and wrote to his father asking if he could switch from theology to mathematics. With permission granted, Riemann spent the academic years 1847–49 in Berlin, which put him right in the thick of the revolution. It also put him at the center of the most exciting mathematical scene in the world. There, famed mathematics professors Jakob Steiner, Carl Jacobi, Lejeune Dirichlet, and

Gotthold Eisenstein were drawing students from all over Europe. Calculus had been applied to new number systems, with stunning results. Daily, it seemed, new functions with amazing properties were discovered. The mathematical discoveries, combined with the social and political unrest of those years, created a rich, exciting, and totally disordered milieu. Riemann mastered complex analysis there and learned a great deal of new mathematics. He was very much taken with the French-educated Dirichlet, whose way of thinking about things conceptually was very much to Riemann's taste.

After several heady months, the revolution stalled. By the fall of 1848, the Prussian aristocrats had regained control in Berlin, the army sweeping the streets. The revolution had failed. Repression returned and hundreds of thousands emigrated to the United States, joining the famine-driven exodus from Ireland. The failure of the Forty-Eighters, as the revolutionaries were called, and the missed rendezvous with democratic unification would have devastating consequences in the twentieth century.

THE EMERGENCE OF THE GERMAN RESEARCH UNIVERSITY

But the same circumstances that gave rise to the revolution played a key role in the emergence of one of the glories of the nineteenth century: the German research university. This was the most effective institution for the creation and transmission of knowledge that had ever appeared. In cultivating excellence and providing venues where individuals as gifted as Riemann could flourish, the research universities underwrote Germany's emergence as the world leader in science and mathematics in the late nineteenth and early twentieth centuries.

Today, we take for granted the role of universities in creating knowledge. This role was not at all clear 150 years ago. During the eighteenth century, science and mathematics were taught primarily in national or regional academies supported by local nobility. And in most regions of Europe, such support was minimal. In Britain, for example, royal financial backing of the Royal Society was practically nonexistent, and mathematics was largely the preserve of gifted amateurs, many of whom occupied university positions in which research was not part of the job description. Royal support of the Prussian Academy was much more extensive, and the academy was the center of science and mathematics in eighteenth-century Germany, although some science was occurring in the universities, especially the newer ones like Göttingen. Research academies in France were concentrated in Paris[65] and were comparatively well-supported financially. The French Revolution had been good for French science. Theology was out. Science was king. And mathematics was queen of the sciences. At the beginning of the nineteenth century, Paris was the intellectual center of the world. Yet French universities were to an even greater extent than elsewhere associated with teaching, and more with rhetoric and the humanities.

No reasonable person would have predicted in the early 1800s that Berlin, or anywhere else in Germany, would soon rival Paris as a scientific or mathematical center. Germany lagged behind Britain, France, and Spain in almost every respect. It was late to unify, late to industrialize, late to urbanize. Napoleon's victory over Prussia at Jena in 1806 seemed to signal that the new century would belong to France. The Congress of Vienna reified the disunity, ensuring that the German states would not get together and that primitive social and economic conditions would continue to prevail.

Paradoxically, however, the disunity and the competition between small states and the relative backwardness of the German economy created incentives for universities to excel. Local princes took pride in their universities and supported them as much as they could. Full professorships were not very numerous and were filled by recommendation from other full professors. Universities vied for the best and most vital professors, making it more likely that individuals would be appointed on the basis of merit. The practice of hiring able professors from other schools provided opportunities for professors to renegotiate their contracts and gave individual professors clout when it came to negotiating conditions of their appointment. Thus, Gauss was able to have Göttingen set up an observatory for him. Moreover, social conditions made academic careers an accessible means of advancement for German youth. And since academic careers were desirable and competitive, professors who were doing cutting-edge research were more likely to attract advanced students. This added to the luster of the university, in turn allowing it to attract even more gifted professors. And so on.

As the nineteenth-century German experience teaches us, the coupling of youth and experience, and research and teaching, makes universities natural and exciting venues for scientific investigation. The mathematician and philosopher Alfred North Whitehead got it exactly right when he wrote, "The justification for a university is that it preserves the connection between knowledge and the zest of life, by uniting the young and the old in the imaginative consideration of learning. . . . The tragedy of the world is that those who are imaginative have but slight experience, and those who are experienced have feeble imaginations. Fools act on imagination without knowledge; pedants act on knowledge without imagination. The task of the university is to weld together imagination and experience."[66] It is no accident that decisive contributions in the sciences and mathematics are often made by the young. Such fields often advance by challenging, instead of building on, received wisdom. While in Germany one could be appointed a full professor at a comparatively young age, the best positions in France were in the national institutes, such as the Academy of Science or the College of France. The French Revolution had stimulated scientific development by clearing out many appointments to these societies, making room for younger individuals. However, appointments were for life, and, as the nineteenth century went on, those to the institutes would too often come only after a scientist's or mathematician's most imaginative years had passed while toiling in positions that left little time for research.

Of course, competition among institutions (and the monarchs who supported them) does not fully account for the emergence of the research university. The French victory under Napoleon in 1806 also contributed indirectly. It caused a lot of soul-searching in the German states, and made room for a number of able individuals outside the university system to play a role in increasing quality. Two such were the von Humboldt brothers. Wilhelm (1767–1835) was the minister of education in Prussia from 1809 and oversaw the development of an education system aimed at providing a rigorous education to all social classes. The University of Berlin opened in 1810 with the explicit aim of being the best in the world. Alexander (1769–1859), a polymath, spent his substantial inheritance on a multiyear trip exploring the Amazon. Trained as geologist, he worked for years on a massive volume detailing his explorations. He corresponded regularly with many of the leading scientists of the time. While in the United States, he visited Thomas Jefferson, then in his second term as president.

The von Humboldts loved the intellectual life in Paris, and were determined to create a similar climate in Berlin. Alexander especially appreciated mathematics, and was instrumental in providing moral and financial support for some of the young mathematicians in Berlin. The University of Berlin had made a very serious attempt to lure Gauss away from Göttingen in the 1820s, but Göttingen countered the offer and managed to keep him. Nonetheless, a vital mathematics department had emerged in the 1830s at the University of Berlin. Alexander lobbied hard to bring Lejeune Dirichlet to Berlin. A controversial but brilliant appointment, Dirichlet brought real intellectual weight and would especially influence Riemann.

Another individual outside the university system who contributed decisively to German mathematics was the Prussian civil servant August Crelle. A civil engineer, he loved mathematics and founded the premier mathematics journal in the world at that time: *Journal für die reine und angewandte Mathematik*.[67] Its title notwithstanding, it is devoted almost exclusively to pure mathematics, and still exists today. Now, as then, it is known as *Crelle's Journal*. Crelle had an almost infallible eye for mathematical talent, and published many now-famous works of then-unknown younger mathematicians. Like Alexander von Humboldt, he personally knew, encouraged, and helped many of them.

By the mid-nineteenth century, the German universities were unrivaled in the depth and quality of their mathematics and science. Like the medieval universities that preceded them, these research centers arose unplanned in a particularly fertile environment. They are the institutions on which the great universities of today are modeled, and they have been a key factor in the explosion in mathematics and science that has fueled our age. The freedom they afforded students and their professors bred excellence and worked especially well for an individual as self-directed as Riemann. When he returned from Berlin in 1849, he took up the study of physics and philosophy, and finished his Ph.D. in 1851 under Gauss's supervision. His dissertation was extraordinary by any measure. Riemann used a principle that he attributed to Dirichlet to make enormous strides in a new field called complex analysis. This field resulted from applying the methods of calculus to the *complex numbers*, numbers obtained by augmenting the real numbers with the square roots of negative numbers.[68] Gauss's report on Riemann's thesis praises its originality and Gauss pushed to have Riemann stay on in Göttingen after his dissertation.

THE HABILITATION

In Germany, as in other countries, the academic career path was the same as that inherited from medieval times in the University of Paris. A student would first do a doctorate, which gave him (there were very few women—social convention and outright prejudice made it very difficult for women to pursue a university career) the right to assist in courses. The graduate would then pursue further work beyond the thesis, submitting the results as a paper (the *Habilitationschift*, or Habilitation paper), and presenting a public, inaugural lecture (the Habilitation lecture). Both the written work and the lecture were judged by a panel of professors. Once the graduate passed his Habilitation, he could be appointed to the university as a professor extraordinarus, which is a bit like an assistant professorship in an American university. The pay was minuscule. The freshly minted professor extraordinarus had the right to offer courses at the university, and would receive a portion of the fees the students paid to take the course. If the courses attracted sufficiently many students, the money was enough to get by. Security and a decent income, however, came only with appointment to one of a small number of full professorship positions (professor ordinarius). Here, the salary was guaranteed and paid by the state.

For his inaugural lecture, Riemann submitted three prospective topics to the faculty. He had done groundbreaking work relating to the first, the second was in an area in which he was expert, and the third was *Über die Hypothesen die der Geometrie zu Grunde liegen* (On the Foundations that underlie geometry). Riemann had not done anything on geometry, although it is clear that he must have been thinking about it for some time. The final choice of topic fell to the supervisor, Gauss, but in practice, the supervisor typically chose the topic with which the candidate was most comfortable. Gauss, however, chose the third, the topic on which Riemann had done the least work. Riemann's friend and colleague Dedekind wrote that Gauss broke with all tradition and chose the third "because he was interested to know how such a young person would handle so difficult a topic."[69]

Lost in time is the tenor of the relation between Gauss and Riemann, so there is no way to know the reason behind Gauss's choice of topic. Since they were at the same place, there are no letters between them documenting any ideas exchanged. The vacuum has been filled with all manner of speculation, including some that Gauss did not fully appreciate Riemann's ability. This is very unlikely, as there is written evidence that Gauss admired Riemann's thesis and had worked to keep Riemann at Göttingen. Riemann, in turn, knew Gauss's work well. Even if Gauss and Riemann had not talked about geometry, the latter surely would have inferred his advisor's interest in the foundations of geometry from Gauss's work on surfaces in three space and his book reviews. Moreover, Riemann was very close to Wilhelm Weber, a talented physicist who was Gauss's son-in-law. One of the seven Göttingen professors dismissed by the reactionary duke, Weber had been reinstated during the revolution of 1848. Gauss trusted Weber and had kept him apprised of his work on the parallel postulate and non-Euclidean geometry.

Altogether, it is inconceivable that Riemann would not have realized that listing a lecture on the foundations of geometry might very well prove irresistible to Gauss. That did not stop him from regretting having listed geometry as a possible topic. "I have become more and more convinced that Gauss has worked on this subject for years, and has talked to some friends (Weber among others) about it," Riemann wrote his father. We all overreach knowing that we will regret the probable consequences. Gauss, for his part, was not so obtuse to choose this topic if he did not think that Riemann would have something interesting to say.

In any case, we have Gauss to thank for insisting on the lecture. Riemann's lecture, so obviously directed at Gauss in particular, started with his great mentor's results, and then turned them on their heads.

WHAT RIEMANN SAID

Riemann first distinguished the notion of space from a geometry, which is an additional structure on a space. He defined a space as consisting of points, and a manifold as a particular type of space which consists of regions in which the points can be named by collections of numbers.

The simplest manifold, often denoted **R**, is the number line—that is, we imagine the real numbers geometrically by thinking of them as corresponding to points on the line. To do this, we draw a line that we imagine going off forever in either direction. We choose a point, to which we assign the number zero, a unit of length (for example, a centimeter, an inch, a fathom, a light-year . . .), and a direction from zero (there are only two) that we decree to be the positive direction. To each positive real number, we associate the point that is exactly that many units of length from zero on the positive side. To each negative number, we associate the point to the left of zero that is exactly as many units of length from zero as its magnitude.[70]

The next simplest manifold is the plane $\mathbf{R}^2$, which we can think of as corresponding to pairs of real numbers. To do this, imagine the page of this book extending infinitely up and down, and also to the right and left. Choose two distinct number lines, which cross at the point corresponding to zero on each. The customary way to draw this is with the first number line running horizontally with the positive direction to the right, and the second running vertically with the positive direction upward. This is just convention, and we could choose other ways to orient the lines. The important thing, however, is that once we have chosen these lines, we associate to each pair of numbers a point on the plane by viewing the first number in the pair as the distance parallel to the first number line, with the sign indicating the direction, and the second number in the pair as the distance parallel to the second line.[71]

Three-space $\mathbf{R}^3$ is the set whose points are *triples* of real numbers (where, as with $\mathbf{R}^2$, the order matters). We picture this geometrically by first laying out three number lines that all cross at the point corresponding to zero on each and which are such that no two are in the same plane. We can't draw this on a sheet of paper (because then the three lines would be on the plane of the sheet of paper), but we can draw two of the lines on the plane of the paper, and imagine the third coming directly out of the paper through the intersection of the two lines and perpendicular to both.[72] We think of the triple (2, 3, –1), say, as corresponding to the point that is two units in the positive direction along the first number line, three units in the positive direction along the second number line, and one unit in the negative direction along the third number line.[73] The triple (2.5, –1, 3) corresponds to the different point 2.5 units in the postive direction along the first number line, one unit in the negative direction along the second, and three units in the postive direction along the first. We imagine three-space extending off infinitely in all directions, each triple of numbers being associated to exactly one point.

Riemann did not stop with numbers, pairs of numbers, and triples of numbers. If n is any positive integer, we think of the set of all ordered n-tuples of real numbers as a space, called *n-space*, and denoted $\mathbf{R}^n$. Its "points" are n-tuples of real numbers and it is n-dimensional because it takes n numbers to specify any point. We can't get more than three independent directions into three-space, because, for example, if we lay out a fourth axis in three-space, we can describe the points along the fourth axis in terms of the triples representing their position with respect to the first three axes. Although we can't really draw a picture of n-space if n is bigger than 3, there is nothing strange or unimaginable about it. If n is 5, say, then five-space is just the set of five-tuples of real numbers. We know what real numbers are, and a five-tuple is just an ordered collection of five of them. So what if we can't plot them? An *n-dimensional manifold* is a set in which the set of points near any given point looks like (that is, is homeomorphic to) a region in *n*-space. Just as when *n* equals to 2 or 3, for any *n*, the simplest manifold of dimension *n* is *n*-space and there are infinitely many other *n*-dimensional manifolds. Riemann even allows infinite-dimensional manifolds.

Euclid had built his geometry on a number of terms for which he offered descriptions, but which were essentially undefined: points, straight lines, planes. True, one could think of a point as a number, the points of a line in one-to-one correspondence with the real numbers, or a plane as in one-to-one correspondence with pairs of numbers, but to get Euclid's theory one had to specify more. Riemann argued that distance was even more fundamental than Euclid's primitive notions and had to be specified independently. Riemann was a master analyst, and the way that he proposed to specify distance in a manifold was as interesting and as fruitful as his other ideas.[74] He observed that once one had a way of measuring speed along any path in a manifold, calculus automatically gives a way of measuring lengths of curves in the manifold, and algebra (and trigonometry) automatically gives a way of measuring angles.[75] We use the word *metric* to mean either a rule for measuring speed along curves or for measuring distance between two points. According to Riemann, these are the same.

Moreover, we define *straight* lines as those that travel along the shortest distance between their points. Such lines are called *geodesics*. To the inhabitants of a space, these are the lines that appear straight. Once we have a notion of straight lines, we can define triangles—these are figures bounded by three geodesic segments. And once we have triangles, we can define curvature. In a two-dimensional manifold, the *curvature* at a point is just a number that measures the deviation from 180 degrees of the sum of the angles of a triangle with vertex at that point. More precisely, it is the measurement of the deviation from 180 degrees as the area of the triangle shrinks. Positive curvature means triangles have more than 180 degrees, negative curvature less than 180 degrees, and zero curvature that the triangles have angle sum exactly equal to 180 degrees.

In a manifold of dimension greater than two, we have many different two-dimensional planes through a point. And we can get different curvatures for geodesic triangles that are tangent to different planes through the point. Curvature is not one number, but a whole collection, one for each pair of directions at a point. (Each pair of directions determines a plane in the space of directions, and this in turn determines a two-dimensional surface traced out by the geodesics tangent to the directions in the plane.) The mathematical gadget for keeping track of the different curvatures in different directions is called the *Riemann curvature tensor*.

In his little book on the differential geometry of surfaces, Gauss had defined the curvature for a surface in three-space, using the behavior of perpendiculars to the surface. His definition did not make sense otherwise because, to define a perpendicular to a point of a surface, we need to have it sitting in three-space. Gauss then spent many pages of computations showing that the curvature could actually be determined by someone living on the surface who could not get off it to draw perpendiculars. In an outrageous stroke of daring, Riemann instead defined the curvature by the property that Gauss had struggled to prove. He defined the space to be *flat*, if and only if every triangle in the space has 180 degrees. This is the case when the curvature in every plane direction is zero. A space, then, is flat if and only if you have what we were calling the Euclidean package. That is, if and only if the Pythagorean theorem holds; if and only if the fifth postulate holds. Three millennia of geometry subsumed in a definition!

To take a particular example, we have just defined two-space $\mathbf{R}^2$ as the set of pairs (x,y) of real numbers. It becomes *Euclidean two-space* if we *define* the distance via the Pythagorean theorem. In this case, we sometimes even use a separate symbol $\mathbf{E}^2$ to make it clear that we have not just two-space but two-space with this particular distance. Likewise we define *Euclidean three-space* $\mathbf{E}^3$ to be three-space (respectively *n*-space) with distance defined by the Pythagorean theorem.[76] One can then show that a curve in Euclidean two- or three-space is a geodesic if and only if it is a straight line in the usual sense. Moreover, since any triangle in Euclidean two-space and three-space has 180 degrees, Euclidean two and Euclidean 3-spaces are flat. There is no reason to stop with dimensions two and three. We define *Euclidean n-space* $\mathbf{E}^n$ to be *n*-space (that is, the set of *n*-tuples of real numbers) with distance defined by a generalization of the Pythagorean theorem.[77] One again, it turns out that all triangles have 180 degrees and thus that, for each *n*, Euclidean space *n*-space is flat.

This shift in point of view completely recast our understanding of geometry, and the relation between Euclidean and non-Euclidean geometry. Once we define a distance, we have straight lines. They are geodesics: the lines that minimize distance between nearby points. There is nothing God-given about

Euclidean geometry. If we *define* distance using the Pythagorean theorem (or so that the Pythagorean theorem is true), then we get the "Euclidean package": parallels are unique, triangles have 180 degrees, similar triangles of arbitrarily large size exist, and so on. A surface, or manifold, with the Euclidean package is flat. By extension, a region in which triangles do not have angle sum equal to 180 degrees, and in particular ones with any non-Euclidean geometry, are not flat. They are curved. What could be more natural? But Riemann's work did far more than potentially reframe geometry. It opened new possibilities for modern science and mathematics, and fundamentally altered the way geometry and topology would develop.

8

Riemann's Legacy

The haunting refrain

All changed, changed utterly:
A terrible beauty is born

of W. B. Yeats's poem "Easter, 1916" captures well the radical shift in point of view that Riemann wrought. Understanding Riemann's point of view is critical to understanding the development of mathematics and science in the twentieth century.

From Riemann's lecture, we know that any surface in ordinary Euclidean three-space has a metric, hence straight lines, hence a geometry. It inherits the metric (that is, a way of measuring distance) from the space in which it sits. In Riemann's terms, a path in the surface is certainly a path in the Euclidean space containing the surface, and therefore we know the speed at any point (because we know the speed at any point of a path in Euclidean space). An equivalent way to determine the distance between two points, without talking about speed, is just to take a measuring tape (think of it as a string that is flexible, but that doesn't stretch—something like dental floss) in the surrounding space, and lay it on the surface between the two points to measure the distance.

SPHERES AND GEODESICS

To further illustrate Riemann's ideas, consider a perfectly round sphere in three-space: that is, the set of all points at a fixed distance from some fixed point. Then the geodesics, or straight lines, on the sphere are the so-called *great circles*—the curves cut out by intersecting the sphere with a plane through the center of the sphere. To see that a great circle traces the shortest distance between any two of its points, take a spherical beach ball or a globe and mark off two points. Now take a string and pull it taut between the two points. The string will lie along a great circle.[78] The longitude lines on our Earth are all great circles, but the only latitude line that is a great circle is the equator.[79] The other latitude lines can be expressed as the intersections of a plane in space and the earth, but the plane does not go through the center of the earth unless one is at the equator. Thus, except for the equator, the latitude lines are not great circles, hence not straight. Note that, as always, by the Earth, we mean the surface of the Earth, so that we are talking about paths constrained to lie on the surface. If we weren't constrained to stay on the surface, the shortest distance between two points would be the straight line in three-space that passes through the mantle and connects the two points.

Through any point, and in any direction, there is a great circle. One way to see this is to treat the given point as a pole, and consider the collection of longitude lines to the opposite pole (figure 28). For example, suppose that we are in Paris and want to go to Boston. Don't think of Paris as being in the middle of the Northern Hemisphere. Think of it as a pole. One of those longitude lines passing through the Paris-pole goes through Boston. That is the direction we want.[80]

Great circle routes seldom appear straight on maps, and routes that appear

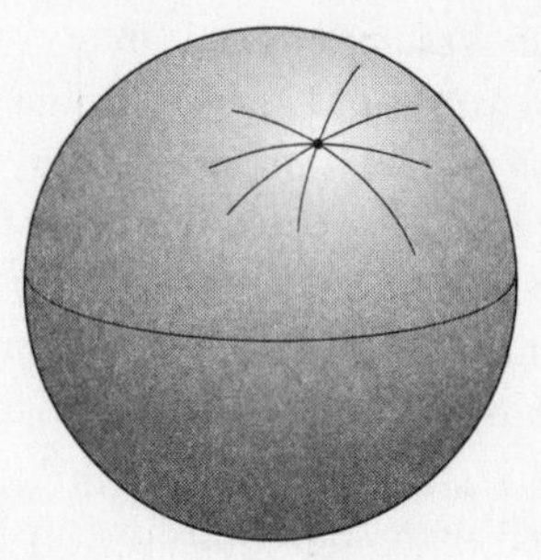

FIGURE 28. *The set of great circles through a point*

straight are seldom geodesics. For example, Beijing and Philadelphia are on the same latitude. If we traveled from one to the other by going along the same latitude, we would go 10,130 miles. The great circle route from one to the other is 6,878 miles, and passes close to the North Pole. This is considerably shorter, and would be what one would perceive as straight if one were piloting a plane. On most world maps, the great circle route would appear to veer north, then head back down. The latitude lines, on the other hand, would look straight. This is because maps of the world on flat paper necessarily distort distance.

Not understanding that great circles are the shortest paths on earth has caused some bizarre arguments. Many religions have preferred directions in which to pray. For example, the ancient biblical tradition among Jews was to pray facing Jerusalem. However, this was usually taken to mean that if one were west of Jerusalem, one should face east. Baha'i tradition states that one should face to Acre to pray. And early Christian practice was to face east. However, these practices were not entirely uniform, and not binding.

Islamic practice is much more precise and prescriptive. The Quran instructs the faithful who are praying to "Turn your face in the direction of the Sacred Mosque: wherever you are, turn your faces in that direction" (Quran 2:144). The direction of prayer, that is, the direction of the Sacred Mosque in Mecca, is called the *qibla*, and mosques are designed to face in its direction. Not only should Muslims pray in the direction of the *qibla*, they should not urinate facing that direction. Since the early Middle Ages, the *qibla* direction has been taken to be the direction that is the great circle direction to Mecca, and many of the greatest scientists in the world, who were Muslims, have worked out methods of determining that direction.

However, opportunities for disagreement abound, and there has been a fair amount of debate in the United States on the *qibla* direction. During the building of the mosque in Washington, D.C., in 1953, the architects consulted the Egyptian Ministry of Works in Cairo for the *qibla* direction and accordingly faced the mosque 56 degrees, 33 minutes, and 15 seconds northeast, a direction slightly north of east. How could this be, visitors to the mosque, the Egyptian ambassador included, wondered? Mecca is slightly south of Washington. A few anxious nights ensued, while the figures were rechecked. The shortest path between Washington and Mecca does indeed run slightly to the north.

This has not set well with some Muslims in the United States, and various directives have appeared decreeing that mosques should face to the southeast instead of the northeast. The exchange has even become fairly acrimonious. One of the causes of the confusion is that a constant-bearing curve (called a *rhumb line*)—that is, one that makes a constant angle with longitudes—is naively assumed by many individuals to be straight. Indeed, rhumb lines are straight when plotted on a Mercator projection of the earth. Since earth has a distinguished point, the pole, to which compasses point, it is relatively easy to navigate such a course. However, such a course is not a geodesic, as we have seen with the latitude connecting Philadelphia and Beijing. In fact, rhumb lines that are not latitudes and longitudes eventually spiral around the poles (think of what happens, say, if one goes northwest continuously).

At any rate, we know that longitudes and the equator are geodesics. Using them, it is easy to construct geodesic triangles: just take two longitudes from the north pole down to the equator, and the segment of the equator connecting then. All such triangles are isosceles with base angles that are right angles. Since the base angles sum to 180 degrees, the triangle has more than 180 degrees. In fact, we can easily construct an equilateral triangle with three right angles and hence an angle sum equal to 270 degrees (figure 29).

It is easy to see that any triangle on the sphere has angle sum greater than 180 degrees. The parallel postulate does not hold. Take a line on the sphere, say a longitude running north–south. Now imagine two great circles crossing it—even if both cross at 90 degrees as in figure 30, they are going to meet (remember: latitude lines are not great circles, and thus are not straight!). The bottom line is that there are no parallel lines on the sphere.

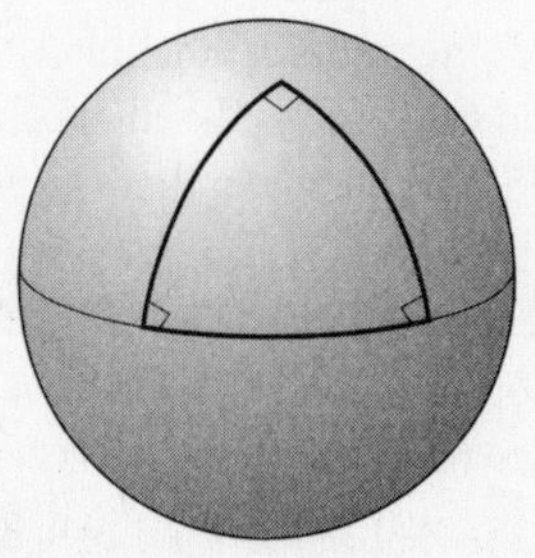

FIGURE 29. *Spherical triangle*

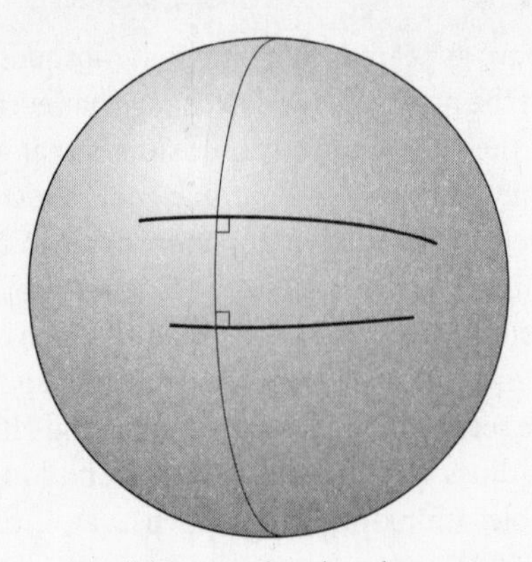

FIGURE 30. *There are no parallel lines on the sphere, because any two lines intersect.*

The geometry on the sphere is not Euclidean. We know this as soon as we have a triangle that has angle sum not equal to 180 degrees. If, as in the case of a round sphere, all triangles on a surface have angle sums greater than 180 degrees, then we say that surface has *positive curvature*. The round sphere is even nicer in that it is perfectly symmetric so that the curvature at each point is the same.[81] On a surface that is not perfectly round, the curvature can vary from point to point, and in fact on Earth it varies, the Earth being slightly flatter (that is, less positively curved) at the poles.

GEOMETRY ON SURFACES

Riemann could show that any space with constant positive curvature is necessarily finite: Geodesics do not extend indefinitely. Great circles necessarily close up. Lobachevsky and Gauss considered another case where there parallel postulate did not hold, namely when there is more than one parallel line through any point off a fixed line. Then it turns out that there must be infinitely many parallel lines through any point off a fixed line. Needless to say, the fifth postulate does not hold: Two straight lines can cross a line, making interior angles less than two right angles, and never meet. In this case, the sum of the angles in a triangle is always less than 180 degrees. Moreover, triangles with larger areas have the smaller angle sums. Any surface with the property

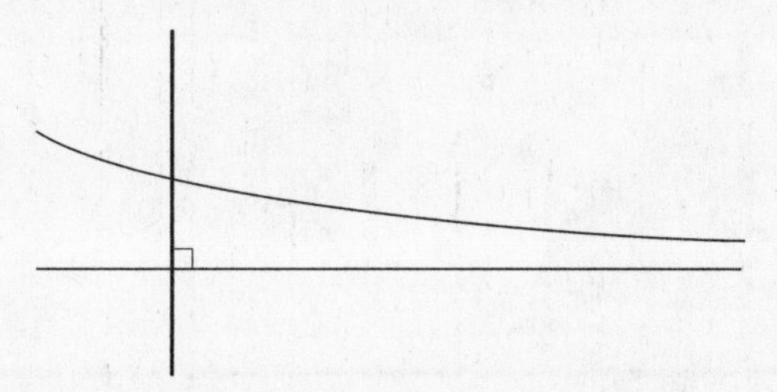

FIGURE 31. *Two parallel straight lines (straight lines on a hyperbolic plane look curved to a Euclidean observer off the plane) crossing a vertical line with interior angles that sum to less than 180 degrees (fifth postulate does not hold).*

that all triangles have angle sum less that 180 degrees is called *negatively curved*.

Surfaces in three-space that are saddle-shaped have negative curvature (equivalently, geodesic triangles on such surfaces have angle sum less than 180 degrees). Conversely, any negatively curved surface in three-space must be saddle-shaped everywhere. In figure 32 we have sketched a triangle whose sides are geodesics. Note that in contrast to the case of a sphere in figure 29, where a triangle whose sides are geodesics (that is, straight lines) seems to bulge out, giving it angle sum greater than 180 degrees, here a triangle whose sides are geodesics on a saddle-shaped surface seems sort of sucked in, making the angle sum less than 180 degrees. Of course, a tiny being living on either surface would view the geodesics as being perfectly straight, and would only be able to tell if the triangle was bulging out or sucked in by measuring the angle sum.

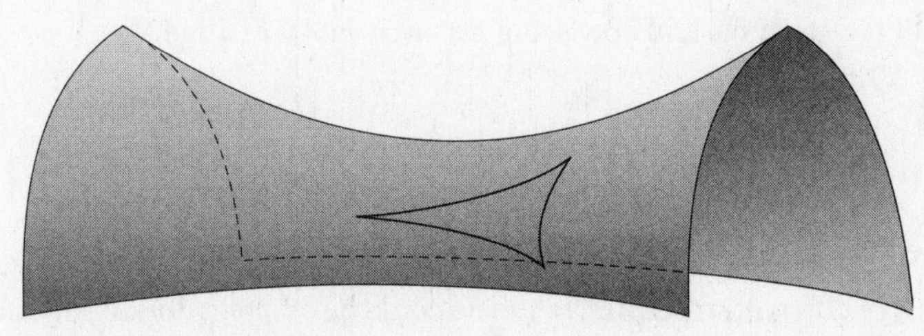

FIGURE 32. *Saddle-shaped surface*

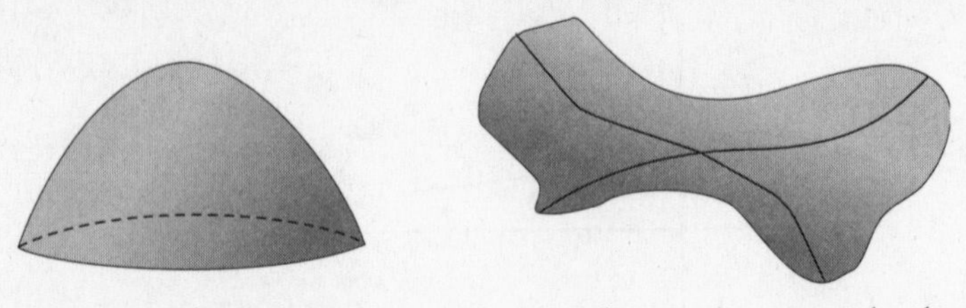

FIGURE 33. *Above left, positive curvature: less circumference and area in a circle. Above right, negative curvature: more circumference and area in a circle*

There is another way to think of curvature. Just as for a plane, on any surface on which distance is defined, we can define a *circle* as the set of points that are a fixed distance from a given point. We have known since Babylonian times that on the plane, the ratio of the length around a circle (that is, the circumference) and the length of a diameter of the circle is a constant, called *pi* and written π, and that the area of the circle is just π times the radius of the circle multiplied by itself. (This is expressed by the formula πr^2 [read "pi r squared"] where the radius r is half the diameter.) On a positively curved surface, the ratio of the circumference of a circle to the length of its diameter is less than π, and the area of the circle is less than that given by the ancient formula. On a negatively curved surface, by contrast, the ratio of the circumference to the diameter is greater than π and the area enclosed by a circle is greater than that given by the formula. In either case, the larger the radius of the circle, the more the area deviates from πr^2.

On a negatively curved surface, if we start out from two very nearby points along geodesics going in directions that appear to be parallel, the lines will diverge, because there is more surface. The opposite is true on a positively curved surface. In his lecture, Riemann did not explicitly mention the work of Lobachevsky and Bolyai, but he surely had it in mind.

DIFFERENT NOTIONS OF EQUIVALENCE

After Riemann, it became clear that the same mathematical object could not only carry different structures, but there could be different notions of equivalence between objects and structures. From one point of view, two objects with different structures might be the same, much as two houses built with the same floor plan but with different materials might be considered the same. From another, they could be as different as night and day.

From a topological point of view, the appropriate notion of equivalence is homeomorphism: two spaces are considered the same if there is a homeomorphism between them. (Recall that a homeomorphism between two spaces is a one-to-one correspondence between them, so that nearby points are mapped to nearby points.) When talking about two-dimensional surfaces, topology is often called "rubber-sheet geometry" because one can visualize a homeomorphism of a rectangle, say, by thinking about what one could do to the rectangle if it were made of an extremely flexible rubber sheet or very stretchy plastic wrap. Mapping a rectangle into a region of the surface with a homeomorphism is like putting clingy plastic wrap onto the region of the surface. Topology studies properties that remain the same under homeomorphisms.[82]

Although Riemann was not primarily interested in topology,[83] he did an enormous amount to enhance our understanding of the topology of surfaces.[84] He introduced the notion of cutting surfaces along closed loops so that what remained became homeomorphic to a rectangle. The smallest number of cuts that will do this is the most important invariant of the surface[85] (and is closely related to the number of tori in the connected sum decomposition mentioned in chapter 3).

Although homeomorphism is the natural notion of equivalence between surfaces (or, more generally, manifolds) from a topological point of view, two surfaces that are the same from a topological point of view can look very different geometric point of view. A round sphere, the surface of a long thin cigar, and the surface of an egg (and, for that matter, all the surfaces depicted in figure 7 in chapter 3) are the same topologically, but different geometrically. As Riemann stressed, a geometry is an extra structure on a manifold. It defines a distance between any two points of a manifold. If a surface is inside a bigger space that has a geometry (such as Euclidean three-space), then it inherits a geometry from the bigger space. One can define different geometries, or distances, on the same manifold. The natural notion of equivalence for manifolds that carry geometries is called an *isometry*. An isometry is a one-to-one map that preserves distances. If two points start out being exactly one-millionth of an inch from each other, then their images under an isometry are exactly a millionth of an inch from each other. If there is an isometry between manifolds endowed with geometries, then the manifolds are called *isometric*.

Isometry is the notion of equivalence that preserves geometries and is different from homeomorphism, which is the notion of equivalence appropriate to topology. Any isometry is a homeomorphism, but not conversely.[86] We have just said that one can visualize a homeomorphism of a rectangle by imagining the rectangle being made of clingy plastic wrap, and the homeomorphism stretching the plastic onto something. To visualize an isometry, which preserves distance, one has to think of dressing, or draping, something with swatches of cloth made of material that is flexible but that won't stretch. If one can fit a rectangle made out of a linen bedsheet onto a part of a surface, then one get an isometry between the rectangle and the part of the surface onto which the bedsheet covers (and that part of the surface is said to be *flat*). Thus, a cylinder is flat, because one can wrap a piece of a sheet perfectly over it. On the other hand, a sphere or the top of one's head is not flat—a part of a bedsheet would not cover it perfectly without wrinkling.

However, one could imaging a superb tailor, or better a weaver, fashioning a spherical piece of cloth that would fit perfectly on the top of your head. Putting the cap on would give an isometry between the cap and the part of your head it covers. The cloth would have to have less area inside a circle of fixed radius than there would be on a bedsheet. Tailors get this effect by putting darts in the material—cutting a piece of fabric out of the fabric, and stitching the result back up (figure 34). Of course, one would need differently curved bits of material to cover the crowns of differently sized heads or differently sized spheres. Any part of the surface of any object that could be covered perfectly, dressed if you will, by this spherical cloth would be isometric to a patch of a sphere (and would have the same positive curvature as the cloth). One would have to crease a piece of positively curved cloth to lay it in a dresser. Think of folding a skullcap.

In contrast to the crown of one's head, the saddle-shaped area on a woman's side above her hip has negative curvature (figure 35), and one can

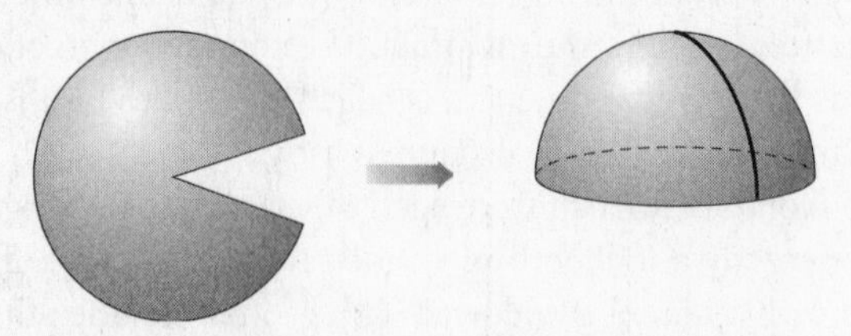

FIGURE 34. *Making positively curved cloth: one needs to cut out darts.*

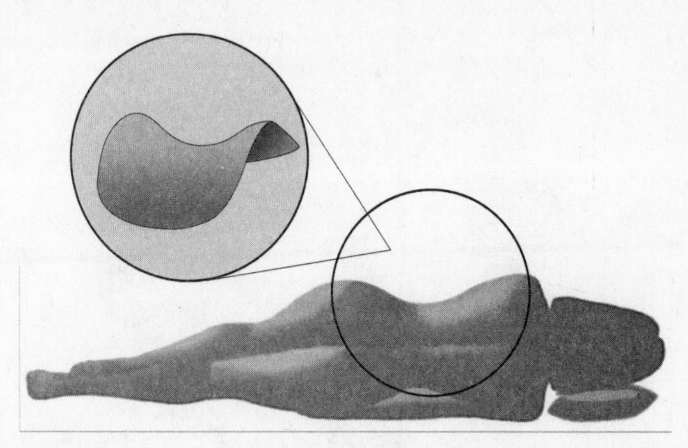

FIGURE 35. *Negatively curved cloth will drape a woman's side.*

imagine cloth that would drape it perfectly. Here, the region inside a circle of a given radius contains more material than the same circle on the plane, and to make the cloth the tailor might start with a flat piece of fabric, make a cut as if he or she were going to make a dart, but instead of stitching the cut edges together, insert an extra piece of fabric or a gusset. Depending on how the area grows with the radius, one gets differently curved pieces.

Negatively curved cloth would have lots of folds if one tried to lay it flat in a dresser (figure 36).

An isometry can therefore be thought of as dressing a part of a surface with a piece of cloth: The salient feature is that the cloth is flexible, but does not stretch.

Incidentally, if one tries to extend a cloth with constant positive curvature in all directions, it would close up, making a sphere. If one imagines extending a cloth with constant negative curvature in all directions, one gets a surface called the *hyperbolic plane*. This surface gets floppier and floppier and one would eventually run out of room in Euclidean three-space.[87] We shall see other ways of thinking about this surface later.

If one takes a polygon on a flat surface and scales it up, the angles stay the same, and the lengths stay in the same proportion. If one takes a polygon on a positively curved surface and scales it up, then the sides of the polygon don't increase as fast as one would expect and the angles between them get bigger (left, figure 37). If one takes the same polygon on a negatively curved surface

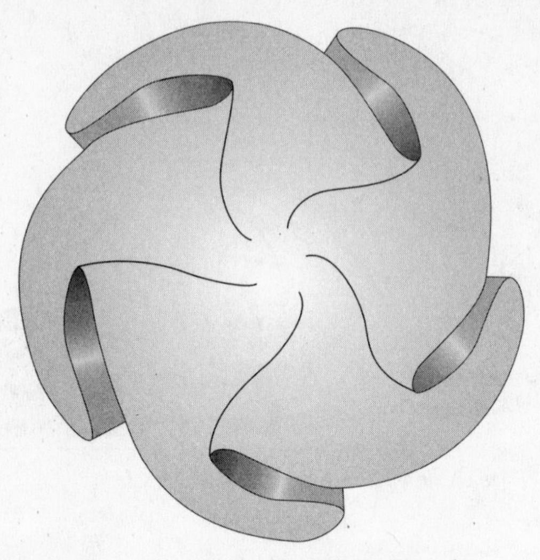

FIGURE 36. *Negatively curved cloth folds when laid on a flat surface.*

and scales it up, then the sides get bigger than one would expect, and the angles get smaller (right, figure 37). If the negatively curved surface goes off to infinity, and the polygon is regular, then one can actually make the angles as close as one may want to zero by scaling the polygon up.

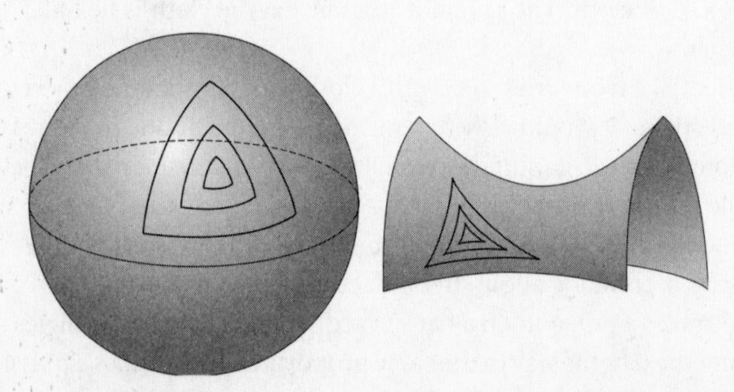

FIGURE 37. *Expanding polygons on a positively curved surface increases the angle sum, and, on a negatively curved surface, decreases it. On a flat surface, the sum remains constant.*

A Summary of the Main Points of Riemann's Lecture

- One needs to distinguish mathematical reality from physical reality. Riemann presumes to talk only about mathematical objects.
- Investigating different mathematical spaces provides possible models for what the universe may be like, and prevents us from being hamstrung by overly narrow preconceptions.
- Continuous spaces can have any dimension, and can even be infinite-dimensional.[88]
- One needs to distinguish between the notion of a space and a space with a geometry. The same space can have different geometries. A geometry is an additional structure on a space. Nowadays, we say that one must distinguish between topology and geometry.
- Manifolds comprise a particularly nice class of spaces—they are spaces that can be charted, in the sense that near every point, their points can be put in one-to-one correspondence with *n*-tuples of numbers. They are the spaces on which one can do calculus.
- A useful way of specifying a geometry on a manifold is to have a method for measuring the speed of an object that is moving along a curve. The speed could differ from point to point. Riemann develops the calculus needed to do this and the consequences. In particular, we can define straight lines and measure angles. Curvature measures the deviation from triangles having 180 degrees.
- A nice class of manifolds and geometries on them is those having constant curvature. These are the only spaces that allow motions of rigid bodies (that is, bodies in which lengths and angles do not change) and are the most symmetric of all spaces.
- It may be that when considered as a whole, our universe looks very different than Euclidean three-space. It may also happen that if we were to magnify and magnify and look at the very small, it would not look like a manifold, but be discrete or something else altogether.

THE EFFECT OF RIEMANN'S LECTURE

Like so much of his work, Riemann's lecture was fundamentally disruptive. Some of the things he talked about were in the air. But he crystallized them, used them to frame and recast what went before in an utterly new manner, and provided new mathematical tools and machinery where needed to implement his concepts and push mathematics in entirely new directions. Once one grasped what he had said, one would never again look at the problem or field under discussion in the same way. He completely changed the terms of discourse. It was as if one were nearly blind, and could suddenly see. Because Riemann changed how we think, it is hard to pinpoint exactly where he influenced others' work—his influence is everywhere, and was the sort that grew, rather than faded, over time as individuals understood and grasped the utility of his ideas. Recognizably modern mathematics begins with Riemann.

The consequences of what Riemann said in his probationary lecture would take many years to understand and to digest. Immediately after his lecture was published, the famous physicist Hermann von Helmholtz announced that he had thought of many-dimensional spaces as well. He argued that the right starting point was to require that one could move solid objects about in the space without having to deform them. Helmholtz could show that this was equivalent to requiring that the space have constant curvature. Constant curvature spaces are geometrically beautiful and attractive, but the extra flexibility that Riemann allows would turn out to be very valuable. The surface of our world is topologically (that is, can be put in continuous one-to-one correspondence with) a sphere, but it is not a constant curvature surface. It is not perfectly round—it is flattened at the poles, and there are bumps (that is, mountains and valleys) on it.

William Clifford, a gifted geometer who first translated Riemann's lecture into English, immediately saw the potential of Riemannian geometry for describing physical phenomena.

> I hold in fact: (1) That small portions of space are of a nature analogous to little hills on a surface which is on average flat. (2) That this property of being curved or distorted is continually being passed on from one portion of space to another after the manner of a wave. (3) That this variation of the

> curvature of space is really what happens in that phenomenon which we call the motion of matter whether ponderable or ethereal. (4) That in this physical world nothing else takes place but this variation, subject, possibly, to the law of continuity.[89]

Sadly, Clifford was like Riemann in another way as well: He died of tuberculosis before reaching age forty. Clifford had neither the time nor the physical expertise to develop his ideas. However, his thoughts foreshadowed a few of the ideas that would be fully developed in general relativity. Riemannian geometry provided the mathematical language needed to express Einstein's ideas. There can be no question that Riemann's deep physical intuition played a vital role in his mathematics, and that he saw clear connections between his geometric ideas and physics.

Riemann did not invent by himself all, or even most of, what we now call Riemannian geometry. He only started it. He elaborated some of his geometric ideas in a paper on heat conduction that he submitted to the Paris Academy, but his health nose-dived after 1862 and he did not live long enough to further develop his geometric ideas. Years of hard, creative work by a great many individuals were needed to elaborate Riemann's ideas.

Although Riemann does not explicitly mention non-Euclidean geometry in his probationary lecture, his work provided the underpinning that would bring non-Euclidean geometry into the mainstream of mathematical thought. It did so by providing a context in which non-Euclidean geometry seemed as natural as Euclidean, and both appeared as special cases in much broader conception of geometry.

In the 1860s an Italian geometer, Eugenio Beltrami (1835–1900), had discovered that geodesic lines on a surface with constant negative curvature behaved just like the lines in the Lobachevskian geometry. He wrote the results up in a now-famous paper, pointing out that it is not necessary to introduce new concepts or ideas to Euclidean geometry to get non-Euclidean geometry. As an example of a surface with constant negative curvature, Beltrami used the *pseudosphere* (figure 38), which is the surface obtained by rotating a curve called a *tractrix* about its asymptotic line.[90] There is a technical problem with the pseudosphere—it has a sharp edge—an inhabitant living on such a surface would get to that edge in finite time. It was not clear whether there were surfaces with constant negative curvature and no edges.

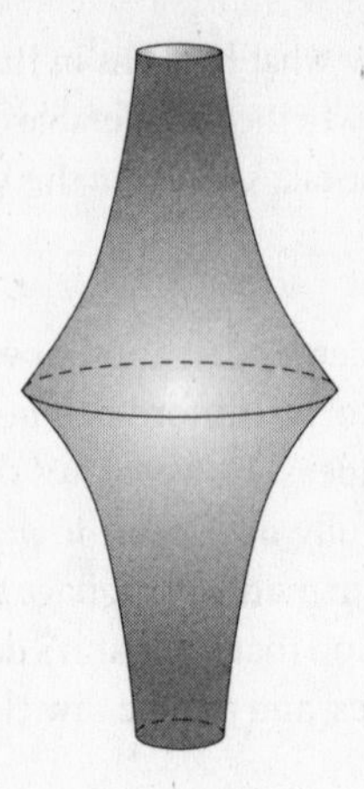

FIGURE 38. *The pseudosphere*

Luigi Cremona (1830–1903), the leading Italian geometer of the time, was uneasy about the draft of the article he saw, and his qualms convinced Beltrami to set the article aside. Beltrami would not have published the article had he not encountered a copy of Riemann's lecture. That lecture made him realize that what he had been saying made perfect sense. Surfaces of constant curvature existed, whether or not they could be put into three-space was irrelevant. Beltrami's paper would have a decisive influence, a few years later, on Poincaré.

Riemann's distinction between physical and mathematical reality is an important one, and is often forgotten. It is enormously liberating to not always have to look over one's shoulder and worry about whether such and such a surface that one can construct explicitly actually can be fit into three-space. Riemann provided mathematicians with an extraordinary playground. We don't need to worry about whether five-dimensional Euclidean space exists—it does. It is just the set of five tuples of real numbers with distance between two points defined by a variant of the Pythagorean formula. What's not to exist? Likewise, it is easy to construct examples of flat tori—we can define them to be pairs, each of which consists of a point of one circle and a point from another, with an appropriate formula for the metric. We can even write it as an explicit subset of Euclidean four-space. You can compute everything. It couldn't be more real.

Some of the mathematical ideas being mooted in German universities at the time spilled out into the popular media. Von Helmholtz's colleague at Leipzig, psychologist and physiologist Gustav Fechner wrote a small story, *Space Has Four Dimensions*, under the pseudonym of Dr. Mises.[91] Fechner worked by analogy, sketching how two-dimensional beings would experience a third dimension. He took time as a fourth dimension. Fechner's analogy would fascinate writers, and in 1884 Edwin Abbot, in England, wrote a lovely little book, part satire of Victorian society, part an extended exercise in geometry, entitled *Flatland: A Romance of Many Dimensions*. The frontispiece reads:

To
The Inhabitants of SPACE IN GENERAL
And H. C. IN PARTICULAR
This Work is Dedicated
By a Humble Native of Flatland
In the Hope that
Even as he was Initiated into the Mysteries
Of THREE Dimensions
Having been previously conversant
With ONLY TWO
So the Citizens of that Celestial Region
May aspire yet higher and higher
To the Secrets of FOUR FIVE OR EVEN SIX Dimensions
Thereby contributing
To the Enlargement of THE IMAGINATION
And the possible Development
Of that most rare and excellent Gift of MODESTY
Among the Superior Races
Of SOLID HUMANITY

This book has been in print ever since, and is still worth reading.[92] It has probably done more to bring the notion of higher dimensions into public consciousness than all the mathematics courses in the twentieth century put together.

THE HUMAN SIDE OF RIEMANN

After all is said and done, there remains an air of profound mystery about Riemann.

He was one of the most daring thinkers of all time, but by all accounts he was almost pathologically shy in social situations. He was comfortable, really, only with his family. How on earth did such a shy individual, deferential and self-effacing to a fault, find the courage to give a lecture that challenged Kant, the most sacrosanct of Enlightenment thinkers? This might have been okay in an audience of mathematicians, but the faculty included philosophers. In an influential recent book, *Good to Great*,[93] American management consultant Jim Collins and a team of researchers identified eight major corporations that consistently outperformed other companies. They found that all were headed by self-effacing CEOs who shared a number of personal traits that Collins dubbed level-five leadership. One of the traits was that no one, except those in the know, had ever heard of the leaders. They were not on magazine covers. They were not flashy. They lived simply and attributed their company's success to luck. They were very modest, attributed their success to others, were determined and extremely focused, and were bold. The combination of modesty with extreme daring and the courage of one's convictions sounds quite a bit like Riemann.

Riemann decided not to publish his probationary lecture—he was too much of a perfectionist and too busy: letters to his brother mention how preoccupied he had become with the connections between mathematics and physics. Riemann would live barely a dozen years beyond his inaugural lecture. He died in 1866, a month shy of his fortieth birthday, on the shores of Lake Maggiore in Italy. His friend and biographer, mathematician Richard Dedekind, portrays his last moments: "The day before his death he worked under a fig tree, his soul filled with joy at the glorious landscape around him. . . . His life ebbed gently away, without strife or death agony . . . He said to his wife 'Kiss our child.' She repeated the Lord's prayer with him; he could no longer speak; . . . she felt his hand grow colder in hers, and with a few last sighs his pure, noble heart had ceased to beat. The gentle mind that had been implanted in him in his father's house remained with him all his life, and he served God faithfully, as his father had, but in a different way."[94]

So many profound ideas, so little time. Riemann would keep generations of mathematicians throughout the world busy. It is worth celebrating the re-

FIGURE 39. Photo of Bernhard Riemann

markable confluence of circumstances that allowed so unlikely an individual as Riemann to flourish. There was the German educational system, the pinnacle of which was the German research university, and there was the remarkable prescience of individuals outside the university system, such as the Humboldt brothers and Crelle, who worked to build excellence. Without Humboldt's pushing, Dirichlet would not have had a position in Germany; and without Dirichlet's influence, Riemann's mathematics would have been far different. Given the intensely human nature of mathematical discovery, we may have to accept the fact that such beautiful mathematics might arise because of a series of accidents. But it is surely sobering to know there are hundreds of young people alive today with Riemann's gifts who will not reach their full potential because of lack of opportunity.

9

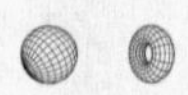

Klein and Poincaré

Poincaré was born in 1854, two weeks before Riemann's fateful probationary lecture. Until he was in his late twenties, Poincaré would know virtually nothing of Riemann. No matter. By that time, Riemann's ideas had begun to interact with those of others and had penetrated almost all areas of mathematics, although in different places at different times and to different extents. Riemann's influence worked both at the level of style and substance. He sought understanding, not just through computation, but by thought, by finding the right concept.[95] His inspiration seemed to well from a deep wordless contemplation of geometry, physics, and philosophy; he transformed every mathematical domain on which his restless intellect alighted. After Riemann, there could be no question that topological and geometric ideas were essential to a deeper understanding of analysis. No longer could there be any serious question that complex numbers were far more than just a convenient shorthand. Higher dimensions and other geometries became central mathematical realities that had a profound bearing on the way we think about our world.

If Riemann's work marks the beginning of modern mathematics, his death in 1866 heralded the dawn of the modern political era. During his lifetime, nationalism arose as a potent force and Europe began to reorganize along lines that we would recognize today. The once-powerful Hapsburg Empire began its slow dissolution. The Italian peninsula was essentially unifed by the mid-1860s. In Germany, the failure of the 1848 revolution had effects that would ultimately prove catastrophic. Conservatives, such as Otto von Bismarck,[96] a law student at Göttingen during the dismissal of the Göttingen Seven, consolidated power in Prussia and created an openly militaristic state that had begun preparing for war with France as early as 1850. The French tactic of preventing German unification at any cost backfired on them and only increased popular sentiment among Germans for unification.

Prussia decisively bested Austria in the Austro-Prussian war of 1866 and removed it from the German Federation. This allowed the forcible realignment of a number of German states with Prussia. When Riemann left for Italy for the last time, Prussia had just invaded Hanover, annexing it. Bismarck, then prime minister, manipulated France into declaring war on Prussia in 1870. The war, now known as the Franco-Prussian War, was short, bloody, and brutal. The Prussians routed the overconfident, ill-prepared French in the battle of Sedan. Emperor Napoleon III was taken prisoner and forced to abdicate. Paris was besieged. Nearly 200,000 soldiers were killed, 80 percent of them French, and another quarter million wounded.

The perception that France was the aggressor had unified Germany. As part of the so-called peace proceedings, the new German empire was proclaimed at Versailles. King Wilhelm IV of Prussia became Kaiser Wilhelm I of the German Reich and the goal of the 1848 revolutionaries was finally achieved, as Bismarck remarked with satisfaction, "forged by blood and iron." The price was the disastrous first half of the twentieth century.[97]

German self-confidence in all things grew. Prussia, which had been lenient with the defeated Austrians, exacted heavy reparation payments from the French that fueled a German economic boom. The rise of the German research university had professionalized the professoriate, and increased the intellectual distance both between academic disciplines and between each discipline and the educated sectors of the populace. The Prussian minister of education, Friedrich Althoff (1839–1908), was a gifted administrator with strong academic training intent on building up German mathematics.

To be sure, there were able mathematicians working elsewhere. Rising wealth and nationalist movements had been good to mathematics. In Italy, a number of very strong geometers were emerging. Paris was still the intellectual capital of the world. British universities continued to be marked not so much by schools, but by the odd gifted individual. The same was true of Scandinavia. Russia had a strong tradition of mathematics that would continue to strengthen. The United States was just emerging from a disastrous civil war, and was experiencing the stirrings of interest in first-class research.

Nothing, however, could rival Germany. The German research establishment was in full swing, now conscious of its own excellence. The combination of competition, a visionary central administration, and a professional professoriate focused on research produced exceptionally high-quality work. The caliber of science in German universities left other places in the dust. The mathematics departments at almost every university—Leipzig, Halle, Konigsberg, Bonn, Erlangen to name a few—were staffed by professors who are still well remembered today. The two greatest universities were Berlin and Göttingen. Berlin had the advantage of a number of schools in easy reach, and a great city. But Göttingen, the university of Gauss, Dirichlet, and Riemann, thrived and outperformed all expectations.[98] Under the visionary and administratively gifted Felix Klein—and those he hired, such as David Hilbert—and under Klein's successor Richard Courant, Göttingen achieved an almost legendary status.[99] Klein, especially, saw the development of mathematics in the spirit of Riemann as an almost sacred trust woven into the very fabric of the University.

FELIX KLEIN

Felix Klein, son of the secretary to the head of the Prussian government, was born in April 1849, as the German revolution was being beaten back. Tall, handsome, well connected, and socially adept, he was a natural leader and quickly pegged as one of the most promising mathematicians of his era. He was in Paris when hostilities broke out in 1870. He hurried back and joined the Prussian war effort as a medical orderly. In the army, he served with future Prussian education minister Althoff, and the two men thought very highly of one other. Following a brief stint at Göttingen after the war, he was appointed full professor at the University of Erlangen at the almost unheard-of age of twenty-three. He moved to nearby Munich three years later in 1875, married, took a position in Leipzig in 1880, and returned to Göttingen in 1886.

Like Riemann, Klein worked in function theory, and had strong interests in geometry and topology. As part of the installation ceremonies on his accession to a full professorship, he had to outline a program of study. That outline, subsequently known as the Erlangen program,[100] laid out a groundbreaking vision of geometry with a different emphasis than Riemann's. One of Riemann's more digestible notions was that one needs to define what one means by a straight line, and how one defines it matters. This had, in fact, been pointed out over a thousand years earlier by Proclus in his commentary on Euclid. Riemann's identification of straight lines with geodesics is the only reasonable definition from the point of view of the inhabitants in a space: lines that are not geodesics would not look straight to them. Even the Greeks in Alexandria knew as much (and they knew that parallel geodesics do not exist in some spaces, and exist but might not be unique in others). However, many felt that Riemann's definition of geometry was too inclusive: If a space has a metric, there is a geodesic through every point in every direction. Shouldn't geometry be special, symmetric, beautiful? Saying that a round sphere has a geometry was one thing, but saying the same of a lumpy potato-shaped blob is another.

For Klein, geometries reflected symmetries; and geometric objects, and lines in particular, were those that remained the same under a prescribed set of transformations. This view had a nice link to *algebra,* in the sense that mathematicians use the word, namely as the study of sets that have one or more operations.[101] A set with a single operation that obeys certain axioms is called a *group.* Sets of one-to-one transformations of a space onto itself have a natural operation: if we do one transformation, and then another, we get a third that can be viewed as the *product* (or composition) of the two transformations. Likewise, the operation that undoes a transformation can be thought of as another transformation, called the *inverse* of the original transformation.

If the set is such that the inverse of any transformation in the set and the product of any two transformations in the set again belong to the set, the set is a group, and it is precisely such sets that are of most interest. Klein's Erlangen program championed the notion that geometry was the study of objects that were invariants of groups of transformations. A different group of transformations on a space gives a different geometry. The groups that come up in the simplest geometries were first studied by Klein's friend Sophus Lie. Spaces that are maximally symmetric with respect to the simplest of these groups turned out to be connected to spaces in which the curvature is constant. It was all very neat.

Klein's and Riemann's viewpoints were not the only ones, of course. There were other mathematicians, notably Hilbert, who felt that one should not need to define points or straight lines, but that there should be a reasonably simple set of axioms that characterized Euclidean geometry. One can think of Riemann's geometry more algebraically, but the algebraic objects had to wait until the second half of the twentieth century.[102]

FIGURE 40. *Felix Klein*

By 1880, Klein had the world by the tail. A virtuoso teacher and a masterful lecturer, he had attracted a large number of talented students to Leipzig. He extended Riemann's work on function theory, investigating functions that were invariant under groups of motions of the complex plane. Klein could not have helped imagining himself as Riemann's intellectual successor. Then in early 1881, Klein saw three short notes by Henri Poincaré, entitled *On Fuchsian Functions,* in the proceedings of the French Academy of Sciences.[103] Who was this guy? A Frenchman? In remote Caen, of all places. Why hadn't Klein heard of him? How did Poincaré get these results? And what was he doing naming functions that deeply interested Klein after Lazans Fuchs (1833–1902), a professor at Heidelberg, who was not nearly as accomplished as some of Klein's closer colleagues?

Klein immediately wrote to the upstart. Nothing afterward would ever be quite the same for Klein: never again would he shine so brightly in his own world. For he had discovered in Poincaré the true intellectual heir to Riemann. Ironically, that heir knew almost nothing of Riemann's work, and he was decidedly not German.

HENRI POINCARÉ

In contrast to Riemann's, Poincaré's family was well off and very accomplished. Henri was born in his grandfather's house,[104] a converted hotel in the heart of the historic French city of Nancy. The house was four stories high, with a central staircase and open central court. It was one block down a road from the Craffe Gate, a monumental fourteenth-century arch through which the local nobility, the dukes of Lorraine, passed before receiving the homage of their duchy. Across the street from the Poincaré home was the church where the dukes were buried. A few short blocks down the street and one to the left was the opulent Place Stanislas, the beautiful Baroque square commissioned by Stanislas Leszczynski, deposed king of Poland and stepfather of the French king Louis XV. Poincaré's father, Léon, was a physician with an appointment in medical school of the University of Nancy. Léon's brother, Antoni, held a succession of very highly placed technical civil servant jobs (he was, among other things, the head of railroads in the Paris region and of the rural French waterworks system). Antoni's sons, Poincaré's first cousins, would attain even higher office: Raymond Poincaré would become president of the French republic,[105] and his brother Lucien, the director of all of secondary, then postsecondary education in France, then vice rector of the University of Paris.

Despite his distinguished bourgeois background, Poincaré was not spared the perils of the age. A case of diphtheria when he was five years old left him unable to walk and speak for nine months. Thereafter his health was delicate, which may account for the slight reserve that one detects between Poincaré and his physically active father. Poincaré's early education, and that of his sister, Aline, was in the hands of his mother, and can only be described as idyllic. The family had outings twice a week with up to twenty people, and Poincaré had access to a great many thoughtful adults. Family vacations to Frankfurt, the world's fair in Paris, the Alps, and London were a part of his childhood, as were long summer and Easter stays at his maternal grandparents' estate at Annecy. Aline cites the latter as among the happiest times of their childhood.

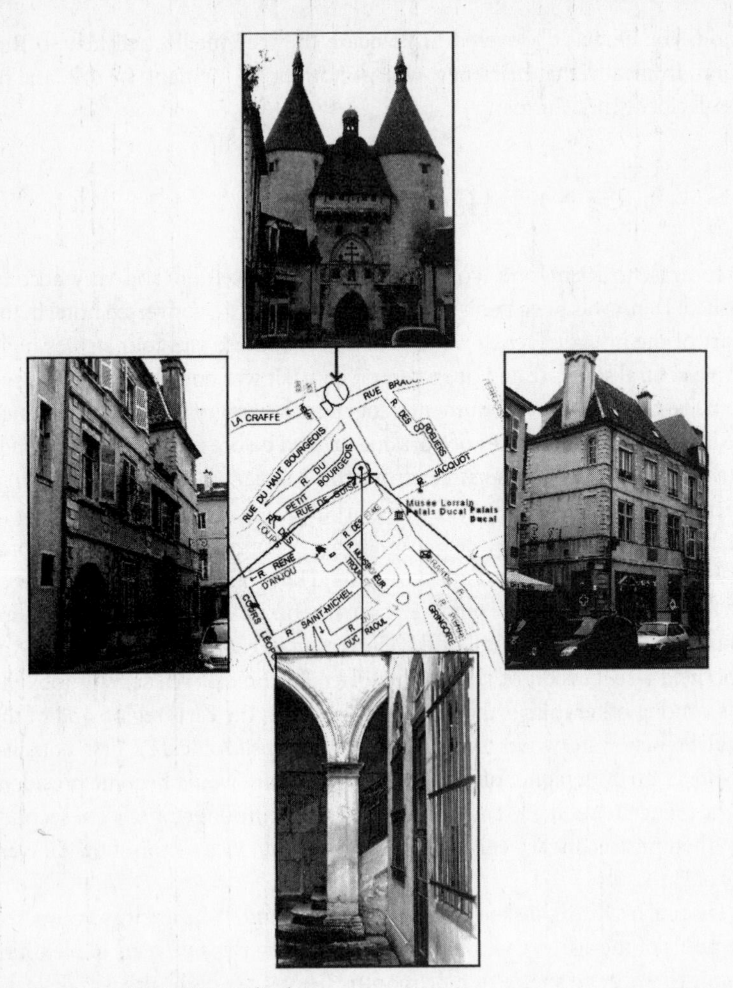

FIGURE 41. *Where Poincaré lived*

His eulogist Gaston Darboux tells of the imaginative games they played at Annecy, elaborating an imaginary constitution for a make-belief empire.

From all accounts, the gentle, abstracted Poincaré seems to have been a source of bemusement to his classmates and teachers. He was ambidextrous, very short-sighted, and his hand-eye coordination was challenged at best, probably as a result of his early bout with paralysis. Yet, except for drawing, school was a breeze for him, and he excelled in all subjects. One of his early teachers told his mother that he would be a great mathematician, writing that Poincaré was a "monster of mathematics." His sister describes him as being buried in books, but never seeming to work. According to her, his mother shielded him from the distractions of day-to-day life. By the time he had arrived home after school, he had done his homework in his head. Years later, classmates would recount the sight of Poincaré dragging out of his pocket an incredibly mangled sheet of paper that was his homework. He never seemed to take notes, and often appeared far away and lost in his own thoughts. He had a virtually photographic memory, easily recalling the exact page and placement of material he had read years before.

Poincaré was even-tempered and good-natured. "He was very balanced. He never showed any anger, any emotion, any passion of any sort. The deepest sentiments were those that he hid most carefully"[106] wrote his sister. "In his judgments of others," she went on, "he avoided all exaggeration. He refused to say that someone was very good, or very bad, because he did not believe in absolutes, particularly in moral matters."[107]

The Franco-Prussian war would play a decisive role in Poincaré's late teenage years. Barely fifty miles from the present German border, the town of Nancy was in the heart of the battlefields. Darboux, who was the head of the French Academy of Science and a great geometer in his own right, describes Poincaré's horrified reaction to the destruction of his mother's family estate outside of Nancy, and quotes a passage from a family friend describing how the young Poincaré helped his father tend to the wounded.[108] Nancy fell to the Prussians, and was occupied from 1870 to 1873, during which one of the high German administrators was billeted with the Poincaré household. It was during this time that Poincaré learned German.[109]

After the war, Nancy returned to French hands. Its sister city, the heavily industrial Metz, thirty miles up the Moselle River, remained in German control until after the First World War and became a part of the German administrative region of Alsace-Lorraine. Nancy, now on the frontier, flourished and developed into a lively border town flooded with artists and refugees fleeing Alsace and the policy of Germanization. The most Roman city in Eastern France, late nineteenth-century Nancy was a beacon of French culture and civilization. Industry, arts, and culture blossomed, helping give birth to the Art Nouveau movement, which spread west from Nancy to Paris and from there throughout France.

One of the enduring legacies of Napoleon was the French system of *grandes écoles*, the elite schools that train the country's top technocratic and managerial students even today.[110] Biographies of French mathematicians often begin with awed accounts of how well they did on the entrance tests, and how they fared on various national exams and competitions. Poincaré was no exception. He obtained first prize in several national competitions, and was among the highest-ranking applicants to the École Polytechnique and the École Normale Supérieure in Paris, schools especially famous for the quality of their mathematics.

Poincaré entered the École Polytechnique in 1873, graduating in second place (on account of his slightly less than average performance in physical education and art) in 1875. He then enrolled in the École des Mines, the grande école devoted to the oldest, and therefore most prestigious, branch of engineering. On graduating, he embarked on a short but distinguished career as a mining engineer, during which time he finished his Ph.D. thesis under Charles Hermite.

Poincaré's thesis examiners were not wholly enthusiastic about his thesis.[111] They acknowledged the wealth and originality of results, but were unhappy with Poincaré's somewhat sloppy presentation of some of them, especially toward the end. Proofs contained gaps, and the manuscript showed signs of having been hastily written. The head of the thesis committee, Darboux, recalled urging Poincaré to further develop some of his ideas and to write them up more definitively. Poincaré dutifully fixed any errors the committee mentioned, but demurred on further developing the ideas, saying that he had too many other things to think about. As Darboux admiringly allowed, he had.

With Ph.D. in hand, Poincaré took an assistant professor position at the University of Caen in December 1879. Service in the provinces was the typical beginning of an academic career in France. Caen is a small city that is the capital of the Calvados region of Normandy, a region famous for its glorious apple brandy and extraordinary cheeses. Known for its Norman buildings, and especially William the Conqueror's massive fortress dating to 1060, Caen is a mere ten miles from the Normandy beaches. The university[112] was founded in 1432 by King Henry VI of England, and would be totally destroyed by aerial bombing associated with the Normandy landings in 1944.

Fatefully, the French Academy of Science had announced in 1878 a prize competition "to significantly improve some aspect of the theory of linear

FIGURE 42. *Henri Poincaré*

differential equations in a single independent variable." Poincaré had been quite taken with some work of Lazarus Fuchs, a coauthor of his thesis advisor, on a class of functions of a complex variable that arose in connection with solutions of differential equations. Poincaré had been wondering if analogous functions existed in other contexts. At the end of May 1880, Poincaré sent off his essay in response to the competition. Shortly thereafter he realized that the functions are connected to non-Euclidean geometry.

"For a fortnight," he wrote, "I had been attempting to prove that there could not be any function analogous to what I have since called Fuchsian functions. I was at that time very ignorant."[113] He went on to describe how, after a sleepless night induced by a late cup of strong coffee, he discovered, contrary to what he had believed possible, a large class of such functions. He then found a convenient way to represent these functions. This was a breakthrough of the first order. However, an even bigger surprise was to come.

> I then left Caen, where I was living at the time, to take part in a geological conference organized by the School of Mines. The incidents of the journey made me forget my mathematical work. When we arrived at Coutances, we took a break to go for a drive and, the instant I put my foot on the step, the idea came to me, though nothing in my former thoughts seemed to have prepared me for it, that the transformations I had used to define Fuchsian functions were identical to those of non-Euclidean geometry. I made no verification, and had no time to do so, since I took up the conversation again as soon as I sat down, but I felt an instant and complete certainty. On returning to Caen, I verified the result at my leisure to satisfy my conscience.[114]

This was even more sensational. But it was not all. Poincaré turned to some arithmetical questions without suspecting that they had any connection with what he had already done.

> Disgusted at my lack of success, I went to spend a few days at the seaside and think of other things. One day, while walking along the cliffs, the idea came to me, again with the same characteristics of brevity, suddenness, and immediate certainty, that the transformations of indefinite ternary forms were identical with those of non-Euclidean geometry.[115]

Back in Caen, Poincaré realized that this implied the existence of yet another whole class of Fuchsian functions. He had stumbled into a mathematical wonderland. Poincaré sent three supplements to his essay to the Academy of Sciences in Paris carefully outlining the connection with non-Euclidean geometry.[116] He was first nominated for candidacy in the Academy of Science in late 1880, and in March 1881 the academy awarded Poincaré a very high honorable mention, although not the first prize.

Poincaré's discovery allowed him to make rapid progress, and he announced his results in two short papers in February 1881, and a third on April 4. These were the papers that Klein noticed.[117] "*Sehr Geehrter Herr!*" (the inimitable formal German formulation of "Dear Sir") began Herr Prof. Dr. Klein's letter to Poincaré, dated June 12, 1881:

> Your three notes in *Comptes Rendus* which I first learned of yesterday, and then only fleetingly, are so closely related to considerations and efforts

> with which I have busied myself these last years, that I felt that I had to write to you. I would like to refer first to some different work that I published in Volumes XIV, XV, XVII of *Mathematische Annalen* on elliptic functions. Elliptic modular functions, of course, are only a special case of the dependency relation you consider, but a closer comparison will show you that I very probably had the general criteria.[118]

Although Klein was barely five years older than Poincaré, he was a fully established German full professor with enormous standing. Klein knew that Poincaré would recognize his name. He carefully documented what had appeared when, making it clear that he had other results that he had not published but about which he had talked to others. The priority claim out of the way, Klein expressed the hope that his letter might be first in an extended correspondence, dropped some names, disingenuously remarked that he would not have a lot of time to think about the matters on which he wrote until a semester later, and apologetically explained that his French was not as good as it should be.

Poincaré was not intimidated. "Monsieur," he replied virtually immediately (June 15).

> Your letter shows me that you have glimpsed before me some of the results that I have obtained in the theory of Fuchsian functions. I am not at all astonished; because I know how well-versed you are in non-Euclidean geometry which is the veritable key to the problem that occupies us.

Poincaré's use of the verbs *glimpse* in reference to Klein's results and *obtain* in reference to his own is subtle, but hardly accidental.[119] Try switching them if you are not convinced. Consciously or not, Poincaré had hit upon the one thing that Klein, of all people, should have discovered, but had not. Klein had not made the connection with non-Euclidean geometry. The problem is one "that occupies us." Poincaré was not ceding the field to Klein. He stated that he would of course acknowledge Klein's priority. Then came a zinger. "You speak of the 'elliptical modular functions.'[120] Why the plural? If the modular function is the square of the modulus expressed as a function of the periods, then there is only one. The expression 'modular function' must mean something else."

It did, of course, mean something else, but Poincaré's German was good enough to notice the plural, and his mathematics was rock solid. Four more questions followed, hard jabs each. "What do you mean by algebraic functions that are representable by modular functions? Also, what is the Theory of the Fundamental Polygon?" Poincaré was not afraid to ask questions and to challenge jargon. "Have you found all circular polygons[121] that give rise to a discontinuous group? Have you shown the existence of functions that correspond to each discontinuous group?" These were, of course, the heart of the matter. It was these latter that Poincaré did not expect to exist, but had discovered after the black coffee. Poincaré closed by saying that he had passed Klein's remark on to Picard, one of the influential French mathematicians whom Klein had mentioned, and by telling Klein that he would be delighted to stay in contact. And, oh yes, "I've taken the liberty of writing to you in French because you told me that you know this language." Perfect. Not quite what Klein said, but a reasonable reading of it.

So began a correspondence that was as fascinating and complex as it was ironic. The letters cut across all sorts of intellectual and emotional fault lines. From outside Germany, Klein epitomized the cultured German elite. Self-assured, handsome, highly educated, and married to Hegel's granddaughter, he had all the perquisites of a German professor with a devoted cadre of students. Within Germany, however, there was a split between the school of analysis typified by the great and influential German mathematician Karl Weierstrass, and the proponents of more geometric methods associated with Riemann. Klein had identified himself, and his students, with the latter, and thereby contributed to widening the rift—for Klein's enthusiasm was the sort that divides as much as it unifies. The rift had consequences outside Germany as well. In France, Poincaré's supervisor Charles Hermite greatly admired Weierstrass but detested anything to do with geometry, and never bothered to learn any of Riemann's methods. As a result, Poincaré's education was rather deficient. Fuchs, who had done some significant work prior to the war with Hermite, was definitely in the Weierstrass camp and very old-fashioned.

So one irony is that Poincaré, who knew almost nothing of what Riemann had done, had begun to re-create and rediscover some of Riemann's results. However, he named his new functions after Fuchs, seemingly aligning himself with the Weierstrass vision. Urbanely at first, then more insistently, Klein suggested that Poincaré rename Fuchsian functions to acknowledge contributions of Schwartz and others. He refused. A second irony is that Klein had carefully studied Riemann's work and the work that grew from it. Poincaré's stunning ignorance gave Klein a huge head start. But Poincaré, who was truly gifted, immediately mastered anything Klein so much as hinted at. The correspondence directed him to Riemann's work.

Another set of ironies arises from the matter of nationality. Relations between France and Germany were terrible. Poincaré, by location of his birth, had seen German aggression firsthand and represented what is quintessentially French. The motto *non inultus premor* that he had affixed to his submission to the contest sponsored by the French Academy of Sciences was that of his hometown Nancy, a prickly slogan that translates roughly as "No injury shall go unavenged,"[122] and that acquired an even deeper sting after France's humiliation in 1870. Klein, for his part, had patriotically enlisted in the German army and served in the war. Although not the chauvinist bigot some have made him out to be, he had a weakness for the myths about national and racial intellectual characteristics that would so disfigure twentieth-century Germany. Years later he would talk almost mystically about the Göttingen atmosphere,[123] and he believed that Riemann's thought epitomized teutonic scientific genius. And yet in Poincaré, he found a young Frenchman who was more Riemann than Riemann himself.

The letters went back and forth for a little over a year. They are superficially courteous but highly charged. A lot was at stake for both men. Klein asked Poincaré to submit a brief paper sketching some of his results to *Mathematische Annalen*, then the leading journal and one which Klein edited. Poincaré agreed. This brought to a head their disagreement over names and attributions. Klein added a disclaimer to Poincaré's paper, stating his disagreement with the choice of the terms *Fuchsian* and *Kleinian* (the latter being Poincaré's somewhat puckish response to Klein's earlier protest to using *Fuchsian*). Poincaré insisted that he be allowed to justify his choice of names. Klein obligingly inserted an excerpt from a letter that Poincaré prepared for this purpose, but wrote to the Frenchman that he was not happy about it.

Poincaré's reply of April 4, 1882, seethes with barely repressed anti-Prussian hostility.

> I just received your letter and hasten to reply to you. You tell me that you wish to close a barren debate for the sake of science and I applaud your resolve. I do so knowing that this resolve did not cost you much since in the note you appended to my communication, it was you who had the last word. So far as I am concerned, I did not open this debate and I entered

into it merely to state once, and only once, my opinion which I refuse to suppress. It is not I who will prolong the debate, I will not speak of it again unless compelled, and I hardly see anything that would so compel me. . . .

I hope that this joust with blunted weapons in which we have just engaged over a name will not damage our relationship. In any case, not having taken the slightest umbrage at your attack, I hope that neither will you take offense at me for having defended myself. It would be ridiculous, anyway, to quarrel any longer over a name, *Name ist Schall und Rauch,* and it is all the same to me: you do what you want, and I, for my part, will do what I want.[124]

This was no courtly joust, it was a dirty street-fight with concealed switchblades—and Poincaré was adept at twisting the knife. He jeered at Klein's determination to close the debate using an expression ("it cannot have cost you much") that would have recalled the reparation payments Germany had extracted from France.[125] He slyly used military terms, which could not have done other than echo 1870, and quoted part of Faust's famously impatient reply[126] to the innocent Gretchen in Goethe's *Faust* (*"Gefühl ist alles, Name ist Schall und Rauch"* [Feeling is all, a name is but noise and smoke]), which was derisive on all kinds of levels; he ended with an abrupt shift into spoken, schoolyard bully language ("you do what you want . . . I'll do what I want"). The correspondence continued until September 1882, when it broke off with both Klein and Poincaré individually having made arrangements to publish full accounts of their discoveries.[127]

The rivalry had driven both men to redouble their efforts. Although the scientific benefit was incalculable, the personal cost was frightful. Neither Klein nor Poincaré needed additional stimulus to work harder. Folklore has a graduate student telling Klein that working too late on mathematics in the evening prevented him from sleeping, and Klein muttering that that was what chloral hydrate was for. Overwork is an occupational hazard for mathematicians: The problems are extraordinarily interesting, and it is difficult not to become obsessed. (A lame but revealing joke repeated by first-year graduate students asks why every mathematician needs a lover. The punch line is, so that one can tell the spouse that one is with the lover, and the lover that one is with the spouse, thereby freeing up time to do what one really wants: sit in the office and ponder that all-consuming problem.)

The stress and overwork affected Klein the most. He broke down completely in the fall of 1882 and entered a serious depression that lasted through

1883 and 1884. "My real productive work in theoretical mathematics perished in 1882," he later wrote.[128] He went on to remark, somewhat bitterly, that he left the field clear for Poincaré. Klein was to contribute much to mathematics, but as an expositor and enabler. Never again would he be on the front lines of mathematical discovery. The book he promised in his last letter to Poincaré did not appear until over fifteen years later. Poincaré, on the other hand, published a complete account in five papers in *Acta Mathematica,* a new journal founded by the Swedish mathematician Gustav Mittag-Leffler. But Poincaré also paid a price. He was diagnosed with exhaustion from overwork in summer 1884 and ordered to rest for a month.

Ultimately both men landed on their feet, even as their careers diverged rapidly. Klein received an offer of a position at Göttingen and turned his formidable talent to exposition, teaching, and administration. He made Göttingen into a center for mathematics and physics unrivaled throughout the world. At the outset of their correspondence in 1881, Poincaré was a relative unknown, though recognized by a few as a potential rising star. He had just been married[129] and in October was to receive an entry-level appointment[130] in Paris. By the time the correspondence ended a year later, Poincaré had come to appreciate Riemann's work, and had mastered and exceeded much of the work of both Klein's and Weierstrass's school. The first of the great *Acta* papers appeared in December 1882; and fifth, and final one, not quite two years later. The papers had firmly established the journal and made Poincaré famous. In 1884, he began an entirely new line of study that would in five years revolutionize mathematical physics. Meanwhile, in 1885, he received a coveted chair (in physics) in the Faculty of Sciences of Paris. In 1887, his first child was born and he was elected to the Academy of Science at the almost inconceivably young age of thirty-two.[131]

10

Poincaré's Topological Papers

The intellectual significance of what Poincaré and Klein had accomplished escaped almost no one. They had opened up an entirely new area of mathematics. Non-Euclidean geometry would ever after sit firmly in the mainstream of mathematics. Moreover, and most interestingly, and although the details would take another two decades to fully elaborate, their labors would lead to an incredibly deep, and totally unexpected, connection between topology and geometry in two dimensions. The resulting theorem is one of the most breathtakingly beautiful in mathematics and, for that matter, in human thought. The Poincaré conjecture was still twenty years in the future and appeared to have no connection with this result. But as fate would have it, the then unimaginable analogue of that theorem in dimension three would be inextricably bound up with the Poincaré conjecture.

To see what Poincaré and Klein had discovered, recall that one of the greatest achievements of the nineteenth century was the classification of all topologically different surfaces. What about the classification of surfaces by geometry? At first blush, this seems hopeless. There are too many possibilities. Spheres of different radius are not isometric, nor are spheres with dimples or mountains. However, Poincaré and Klein's work implied that any surface could be given a geometry in which it had a constant curvature (and it followed easily that this geometry had to be unique—if a surface had a flat geometry, we couldn't put a spherical geometry on it).

THE NATURAL GEOMETRY ON A SURFACE

This isn't too surprising for a sphere. We know that a perfectly round sphere is homeomorphic to any sphere and hence that the sphere has spherical geometry. But, what about a torus? Any torus in three-space has regions of positive and negative curvature. Along the outside of the torus, triangles have angle sum more than 180 degrees, so the curvature is positive. On the inside, the torus is saddle-shaped and triangles have angle sum less than 180 degrees, so that the curvature is negative. Now, let's think of the torus as we did in chapter 3, imagining that is the surface obtained from a square by matching its edges to their respective opposite sides. That is, we think of each point on the bottom edge of the paper as the same as the corresponding point on the top edge vertically above it, and each point on the left edge as the same as the point directly to the right.

Let's also take Riemann at his word and specify a geometry on the torus, by saying that we will measure length and angles on the torus by measuring the corresponding lengths and angles on the square.[132] We have to remember that when something goes out one of the sides it comes back in at the corresponding point on the other side. The effect is to say that the straight line segments on the torus are straight line segments on the square, with the proviso that they might go out one edge and come back in the other. Indeed, the shortest line segment joining one point to another might be out one side and back the other (figure 43).[133]

Any triangle on this torus has 180 degrees. Therefore, it is flat and its geometry is Euclidean (but unlike the Euclidean plane, there are finite straight lines and it has a finite area). For this reason, it is called a *flat torus*.

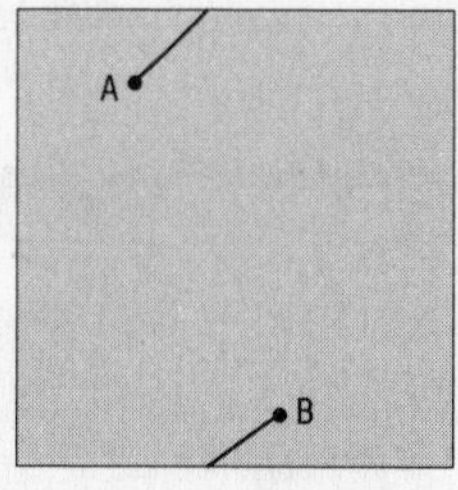

FIGURE 43. *Straight line joining two points on the torus*

The effect of connecting the top and bottom edge of the square is to roll the square into a cylinder. This does not distort distances and can easily be done in three-space. Since it does not distort distances, straight lines and triangles drawn on the square become straight lines on the cylinder. In particular, because its triangles have 180 degrees, the cylinder is flat. We can easily wrap a flat bedsheet onto a cylinder (figure 44).

However, once we try to connect the right and left edges in the same way in three-space, and to pull the circles at each end of the cylinder around so that they join each other, we wind up distorting distances, and we introduce curvature. In going from the top right figure in figure 2 to the bottom right, we have changed distances between some points. It turns out that there is no way to get a torus into three-space without distorting distances and therefore introducing curvature. However, because we can't do it in three-space does not mean that it does not exist. Of course it does! We have a well-defined set. We have a well-defined distance. We don't need anything more. For a mathematician, the flat torus is the more natural one than the one embedded in three-space. After all, three-space itself is a mathematical construct—it doesn't exist outside of mathematics. When we think of an exemplar of a sphere, we think of a round sphere, without any changes in curvature.

In contrast to a regular torus, a two-holed torus has a metric with constant negative curvature. That is, its most natural, most symmetric state has a hyperbolic geometry. The argument that this is the case follows the same line as the previous argument that the natural geometry on a torus is Euclidean

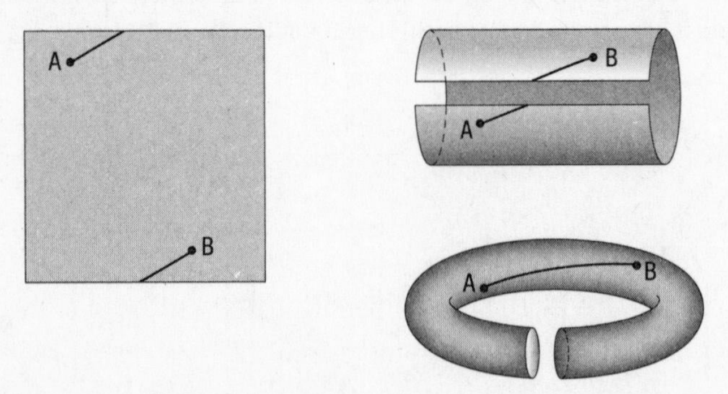

FIGURE 44. *Connecting the top and bottom sides does not distort distance in three-space. Joining left and right sides does.*

(that is, flat), but is a little more complicated. First, observe that the process of cutting up a two-holed torus so it could be unfolded onto a plane (figure 10 in chapter 3) actually shows that you can consider the two-holed torus as an octagon (an octagon is a polygon with eight sides) with alternate sides identified in pairs (figure 44).

We can't do the same thing that we did with the torus and use the lengths and angles of a regular octagon (one whose sides all have the same length) to define a geometry on the two-holed torus. The reason is that, after we match up all the like points, all the vertices of the octagon are identified with a single point, and the eight angles in the octagon come together around that point. But there are 360 degrees around a point (this is the definition of degree), and thus if we wanted the eight angles to fit perfectly around a point, each angle would have to have 45 degrees (since 8 divided into 360 is 45). But it is clear from figure 45 that each of the eight angles in a regular octagon has much more than 45 degrees, so they are not going to fit.[134]

Recall, however, that in hyperbolic geometry triangles have less than 180 degrees and that when we scale up polygons their angle sum gets smaller. This raises the possibility that we could make a regular octagon in a hyperbolic geometry with angles that have 45 degrees. We could then join up all sides and it would fit perfectly around the common vertex. Phrased differently, perhaps we could cut a regular octagon out of hyperbolic cloth in which each angle had 45 degrees—and we could then attach the sides to each other to make a perfectly symmetric double torus. This would allow us to assign the distances in this non-Euclidean octagon to measure distances on the double torus (as in the case of the torus, the shortest distance may be after

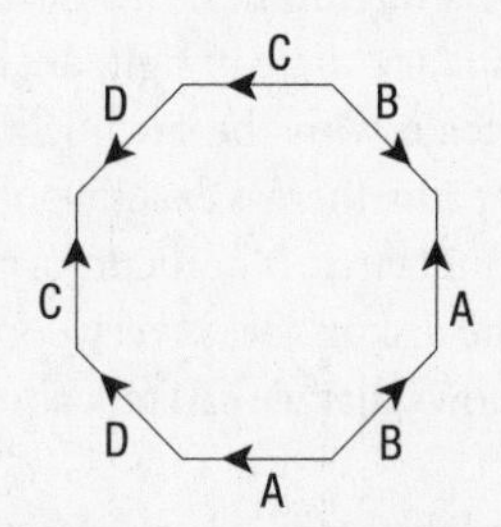

FIGURE 45. *Matching up corresponding points of sides that are labeled with the same letter (in the directions indicated by the arrows) gives a two-holed torus. However, each of the angles is greater than 45 degrees so one can't bring them together around their common point unless one distorts angles (and distance).*

crossing an edge and coming back in). With this metric, the two-holed torus would have constant negative curvature (because the hyperbolic plane does). The region around every point is isometric to the region around any other point. It is perfectly symmetrical.

The insight necessary to establish that the above construction is both possible and meaningful, is exactly the one that occurred to Poincaré upon stepping on the carriage in Coutances. Of course, to work out the details, Poincaré needed a more analytic way to think of a constant curvature geometry than that which we have sketched in chapter 8 using a uniformly woven fabric. Poincaré had worked out different ways of thinking of the geometry considered by Lobachevsky by imagining a piece of the ordinary plane and defining the distance differently from the Euclidean distance, so that from a Euclidean viewpoint the geodesics (that is, the straight lines) appear to be warped. One such way is the *Poincaré disk model* of the hyperbolic plane. Take a disk—that is, the region bounded by a circle in the ordinary plane (not including the circumference). Imagine that distances are defined so that they become larger and larger as we near the boundary, making that boundary infinitely far away. This seems different than a surface made of a cloth that is woven so that every circle contains more area than it would in the Euclidean plane. But the two ways of thinking about the hyperbolic plane are equivalent: In one, we are taking the point of view of a bug living on the surface—every part looks the same, the plane extends to infinity. In the other, we imagine ourselves taking an extra-dimensional view outside the surface and seeing it as a disk in which distances get greater as we approach the boundary.

In the disk model, it turns out that we can define the distance on the disk so that (1) the geodesics are straight lines through the center of the disk and circular arcs that meet boundary circle in right angles; (2) angles between geodesics that intersect coincide with the Euclidean angles, and (3) through every point, and in any direction, there is exactly one geodesic (figure 46). Incidentally, the interior of the unit circle is homeomorphic to the plane, whatever the geometries on each. The disk model is very convenient for computation. Just as important, it also shows that we can talk about non-Euclidean geometry in Euclidean terms.

With this model in hand, let's find a regular octagon in the disk with the metric Poincaré used. On the left, figure 47 shows a regular octagon that is small and for which each angle is greater than 90 degrees. (Remember that straight lines are circles that meet the boundary circle at 90 degrees.) On the

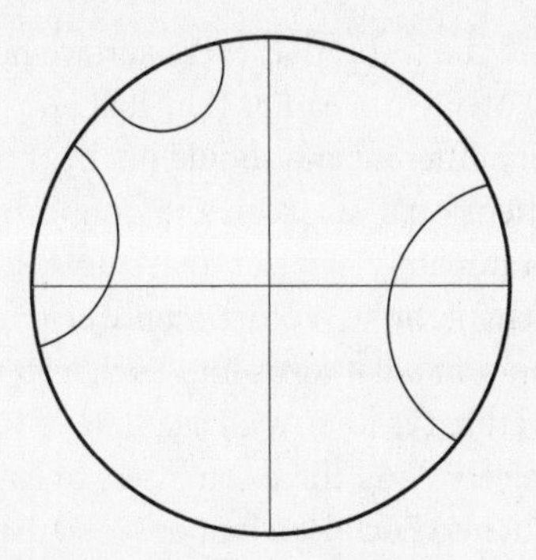

FIGURE 46. *The Poincaré disk model with some geodesics*

right, we display a regular octagon in which every angle is 0 degrees (the vertices are at infinity). If we expand the octagon on the left until it becomes the one at right, the angles get smaller, eventually reaching 0 degrees. For some intermediate octagon, the angles will have to equal exactly 45 degrees, just as if we slow a car traveling faster than 90 miles per hour to a stop; at some instant before it stops, it will be going exactly 45 miles per hour.

This octagon, with all angles equal to 45 degrees, is the one we seek. If we match up the sides in the same way they are matched up in figure 45, we get a two-holed torus. Defining distances as they are defined on the octagon gives the latter a geometry that is negatively curved.

What about three-holed tori, or tori with any number of holes? By an argument similar to the one above, these all can be given metrics with constant negative curvature, so are hyperbolic. So, the natural geometry of "most" surfaces is hyperbolic.[135]

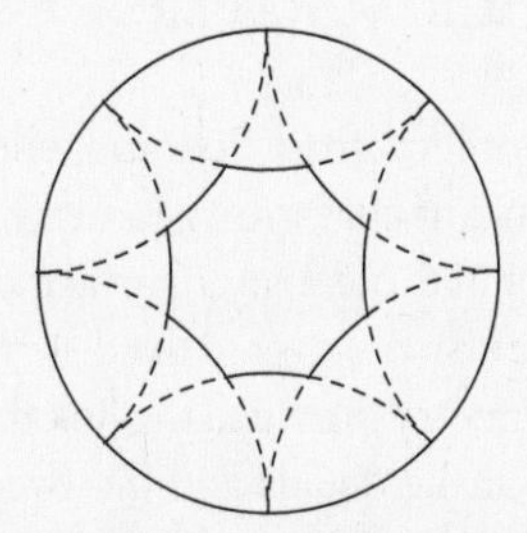

FIGURE 47. *The octagon at left has angles greater than 90 degrees. The octagon at right has angles equal to 0 degrees. As the octagon at left expands, its angles shrink, becoming exactly 45 degrees for some octagon before reaching the one on the right.*

Mathematicians find the result that every surface has a natural geometry irresistibly beautiful. Forty years earlier they had just learned that topology and geometry were very different and should not be conflated. Now, it turned out that, for two-dimensional surfaces, the connection between the two could not be more intimate. The fact that most surfaces had the still-somewhat mysterious hyperbolic geometry only heightened the wonder.

Although, one cannot fit a flat torus into Euclidean three-space (although one with varying curvature fits fine), with a couple of years of college mathematics (a little linear algebra and three semesters of calculus), one can show that the torus can be put into Euclidean four-space so that the metric it inherits is the flat one.[136] Hilbert showed that no surface could be embedded in Euclidean three-space so that it acquires a metric with constant negative curvature. In particular, a hyperbolic two-holed torus does not fit into Euclidean three-space. One could ask if it is possible to fit every surface, and more generally every manifold with a metric in the sense of Riemann, into some *n*-dimensional Euclidean space so that the metric (hence distance) on the manifold is the same as the one it inherits from the ambient Euclidean space? This question is known as the *Riemann embedding problem* and went unsolved for many years. In 1956, in an extraordinary paper that seemed to pull results from nowhere, John Nash proved that the answer to the general embedding problem was yes. The result made him famous, but shortly thereafter he descended into a deep psychosis. His emergence from the grip of schizophrenia and his winning of the Nobel Prize for his contributions to economics are the subject of the book *A Beautiful Mind* and a popular movie with the same title.[137]

THE GREAT TOPOLOGICAL PAPERS

Poincaré had learned of non-Euclidean geometry from Beltrami who first noticed that hyperbolic geometry was a constant curvature geometry in which the curvature was negative. Beltrami pointed out that a surface called the pseudosphere carried such a geometry. He also discovered the disk model and realized that Riemann's conception of geometry provided the link that united the two conceptions. Poincaré worked out the details and discovered more models. Moreover, Poincaré did not stop at two-dimensions. He extended this work to higher-dimensional geometries. Here he is, in his own words, describing his model of three-dimensional hyperbolic space:

> Suppose, for example, a world enclosed in a large sphere and subject to the following laws: The temperature is not uniform; it is greatest at the centre, and gradually decreases as we move towards the circumference of the sphere, where it is absolute zero. The law of this temperature is as follows: If R be the radius of the sphere, and r the distance of the point considered from the centre, the absolute temperature will be proportional to $R^2–r^2$. Further, I shall suppose that in this world all bodies have the same coefficient of dilatation so that the linear dilatation of any body is proportional to its absolute temperature. Finally, I shall suppose that a body transported from one point to another of different temperature is instantaneously in thermal equilibrium with its new environment. There is nothing in these hypotheses either contradictory or unimaginable. A moving object will become smaller and smaller as it approaches the circumference of the sphere. Let us observe, in the first place, that although from the point of view of our ordinary geometry this world is finite, to its inhabitants it will appear infinite. As they approach the surface of the sphere they become colder, and at the same time smaller and smaller. The steps they take are therefore also smaller and smaller, so that they can never reach the boundary of the sphere. If to us geometry is only the study of laws according to which invariable solids move, to these imaginary beings it will be the study of laws of motion *deformed by the differences of temperature* alluded to. . . .
>
> Let us make another hypothesis: suppose that light passes through media of different refractive indices, such that the index of refraction is inversely proportional to $R^2–r^2$. Under these conditions it is clear that the rays of light will no longer be rectilineal, but circular. . . . If they [the beings in such a world] construct a geometry, it will not be like ours, which is the study of movements of our invariable solids; it will be the study of the changes of position which they will have thus distinguished, and will be "non-Euclidean displacements," and *this will be non-Euclidean geometry.* So that beings like ourselves, educated in such a world, will not have the same geometry as ours.[138]

Throughout his life, Poincaré would brush up against complexities that would take nearly a century for others to appreciate. His first such encounter involved the model of hyperbolic three-space that he outlined above. He discovered that the actions of different subgroups of motions on the sphere at infinity were far more complicated than anything mathematicians had previously encountered. A few years later (1887), the king of Norway and Sweden announced a mathematical competition for the best work on the mathematics of the solar system. It turned out that the goal sought was far too ambitious. Poincaré submitted a memoir that won the prize, and then discovered a mistake in it.[139] He had assumed that a certain type of infinitely intricate behavior could not happen, and then he discovered that it could. He had discovered what we now call *chaotic behavior,* and would subsequently seek language and tools to describe and tame what he had found. Nowadays, we use the term *chaos theory* to describe the sometimes unimaginably complex behavior that can result from simple laws of motion or simple rules repeatedly applied.[140]

To make sense of what he was apprehending, Poincaré, like Riemann, realized that topological notions were necessary. In a retrospective of his work in 1901, he wrote:

> A method that allows us to know the qualitative relations in a space of more than three dimensions could, in a certain way, provide services analogous to those that figures provide. This method can only be the topology of more than three dimensions. Despite this, this branch of Science has been to the present little cultivated. After Riemann, Betti introduced some fundamental notions, but Betti has been followed by no one. For my part, all the diverse paths on which I have successively found myself engaged have led to topology. I needed notions of this Science to pursue my studies on curves defined by differential equations, and to extend them to higher order differential equations and, in particular, to the three-body problem. I needed them to study non-uniform functions of two variables. I needed them to study periods of multiple integrals and to apply this study to perturbative expansions. Lastly, I foresaw in topology a means of tackling an important problem in the theory of groups, the study of discrete or finite subgroups of a given continuous group.[141]

What he did not say is that the notions that he required were far beyond those that Riemann needed, and far beyond what was available at the time. Poincaré's contributions to topology were stunning and well worth reading even today. Apart from research announcements, they appeared in six papers beginning in 1895 and ending in 1904. The first appeared in the first volume of the journal of the École Polytechnique and in over a hundred pages laid out in an almost leisurely fashion the foundations of the field.[142] It is an absolute tour de force. He gave several different definitions of manifold: Two are very convenient for analysts, another is more convenient for the creation of examples in low dimensions and forms the basis of what we call geometric topology, and a fourth makes a link with group theory. The latter became crucial for algebraic topology, a field that Poincaré almost single-handedly invented.

Poincaré discussed the work of Betti (whom he also mentions in the extract we quoted above), who was a friend of Riemann and probably the only one to fully understand Riemann's topological ideas. Betti had taken Riemann's idea of cutting a surface along curves and generalized it to cutting a higher-dimensional manifold along a manifold inside it. Poincaré, in turn, generalized Betti's work and the numbers that Betti associated with manifolds. He defined a *submanifold* to be a manifold inside another manifold (like a curve inside a surface, or a curve or surface inside a three-dimensional manifold) and considered two or more submanifolds related if they were the common boundary of yet another submanifold. He reinterpreted the numbers that Betti had considered, nowadays called *Betti numbers,* by introducing equations between submanifolds of a manifold, called *homologies* on a manifold that expressed the relation of bounding within the manifold. The collection of all independent homologies of a fixed dimension formed a group called a *homology group*. The homology groups were conjecturally the same for manifolds that were homeomorphic and once you knew the homology groups you knew the Betti numbers and more.[143]

Poincaré associated a completely new algebraic object with each manifold, which he called the *fundamental group*. It is invariant under homeomorphism and completely revolutionized how we think about manifolds. An element of this group is a loop based at a fixed point in the manifold; that is, a path that leaves the point and returns. More formally, it is a continuous map of the interval into the manifold so that both endpoints map to the distinguished point. Two paths are deemed equivalent if they can be deformed into each other in a continuous manner fixing the starting (and hence the ending) point. We can multiply two paths by going first around one, then around the other. The set of paths turns out to form a group, but one in which the order in which we multiply paths matters. He worked out how we can describe such groups, and came up with a scheme whereby the group can be realized as the set of all words, or strings, we can create with a given set of letters. Equivalence amounts to being able to replace some strings with others, and multiplication to juxtaposing two strings. For any group, the letters we are allowed to use are called *generators,* and the equivalence rules are called *relations.*

Poincaré was particularly concerned with three-dimensional manifolds. These modeled possible shapes that our universe might have. Riemann had earlier considered the three-dimensional sphere. Poincaré was interested in understanding *all* three-dimensional manifolds.

As examples, Poincaré considered the class of three-dimensional manifolds that are obtained by matching up opposite faces of a cube to one another in various ways. He got the three-dimensional torus mentioned in chapter 4, as well as some other manifolds (together with an object that he showed is not a manifold). He showed how to determine whether the objects one obtains are manifolds. He studied their fundamental groups and showed that these manifolds can be thought of as the space one gets if one identifies points in three-space that can be moved into one another by a particular group (one says that the manifold is represented as a *quotient of a group* acting on three-space).

Poincaré was particularly interested in finding a set of invariants that would distinguish different manifolds. (Put differently, since we lived in a universe that is a three-dimensional manifold, how can we tell which manifold it is?) He showed that it is not enough to know the Betti numbers. He produced an infinite family of closed three-dimensional manifolds that are not homeomorphic to one another and showed that one can have nonhomeomorphic manifolds with the same Betti numbers and, in fact, with the same Betti numbers as a sphere. He provided examples of manifolds with finite fundamental groups, and signaled the "seemingly paradoxical" (his words) result, on which he hoped his paper shed some light, that one manifold can have a fundamental group that is much more complicated than another, yet have a smaller first Betti number. He went on to write:

> It would be interesting to investigate the following question.
>
> 1. Being given a group G defined by generators and relations, can it be the fundamental group of a manifold of *n* dimensions?
> 2. How can one form this manifold?
> 3. Are two manifolds with the same dimension and same fundamental group always homeomorphic?
>
> These questions involve difficult studies and long developments. I will not talk about them here.[144]

As the twentieth century dawned, five papers that he called "complements" followed the great 1895 paper. All appeared in top-notch journals.[145] The first complement in 1899 clarified, in response to criticism from the Danish mathematician Poul Heegaard, the definition of Betti numbers. Heegaard gave a counterexample, showing that a theorem, now known as Poincaré duality, could not be true as stated. It turned out that Poincaré's definition differed from Heegaard's, and the difference was crucial in allowing Poincaré duality to work. In the second complement, which appeared in 1900, he discussed *torsion coefficients,* which are refinements of Betti numbers, and extended his duality theorem to them. He revisited his examples of three-dimensional manifolds. The third studied a particular class of algebraic surfaces.[146] These have two complex and hence four real dimensions. Poincaré introduced entirely new ideas, looking at changes in two-dimensional surfaces inside the manifold as one moves around *singular points.* The fourth broadened the study to arbitrary algebraic surfaces. In the fifth and final complement, published in 1904, Poincaré returned to three-dimensional manifolds, with the extra machinery he had in hand.

The simplest three-dimensional manifold is the three-sphere, and it is clear that the very specific case of characterizing it was ever-present in Poincaré's thoughts. In the second complement, he thought he had it. He began by pointing to his two preceding papers. "Nevertheless the question is far from being exhausted, and I will doubtless return to it several times. This time, I restrict myself to certain considerations aimed at simplifying, clarifying and completing results previously obtained."[147] He went on to develop and carefully lay out a procedure to compute the coefficients of torsion that he had defined in the second complement and to sharpen the duality theorem. However, he overreached.

> In order not to prolong this work, I restrict myself to announcing the following theorem whose proof would require some further ground work.
>
> *Every polyhedron which has all its Betti numbers equal to 1 and all its arrays* T_q *bilateral is homeomorphic to the three-dimensional sphere.*[148]

And so begins the tortuous history of the Poincaré conjecture. Poincaré believed that he had characterized the three-dimensional sphere. But the "theorem" he announced and italicized is false. Four years later, he realized this, and devoted the fifth complement to constructing a stunning counterexample. The paper begins with the following words:

> I return again today to this same question [topology], persuaded that one can succeed only by repeated efforts, and that the topic is sufficiently important to merit them. This time I limit myself to the study of certain three-dimensional manifolds, but the methods that I use can doubtless be applied more generally. In passing, I dwell at length on closed curves that one can trace on closed surfaces in ordinary space. The final result that I have in mind is the following. In the second complement, I showed that to characterize a manifold, it does not suffice to know its Betti numbers, but that certain coefficients that I called torsion coefficients play an important role. One could then ask whether a consideration of these coefficients suffices; that is, if all the Betti numbers and all the torsion coefficients of a manifold are equal to 1, is that manifold homeomorphic to a three-dimensional sphere? Or is it necessary, before affirming that the manifold is such, to study its fundamental group? We can now respond to these questions; I have constructed, in effect, an example of a manifold all of whose Betti numbers and torsion coefficients are equal to 1, but that is not homeomorphic to the three-sphere.[149]

In other words, Poincaré reminds the reader that if one knew all the Betti numbers of the universe, one still wouldn't know its shape. Now, he asks whether, if one knew the Betti numbers and the torsion coefficients, one could determine the manifold? No, he says. He has an example of a manifold that has the same Betti numbers and the same torsion coefficients as the three-dimensional sphere, but which is not the three-dimensional sphere. This was surprising; In the second complement, Poincaré had stated that he didn't think this was possible.

What was Poincaré's example? In the fifth complement, he described it as two solid two-holed tori glued together in an appropriate manner (a *Heegaard diagram*), and carefully described a pair of paths that do not shrink to a point. The manifold has since been known as *Poincaré's dodecahedral space* (figure 48), and is nowadays usually described as the three-manifold that we get by gluing together opposite faces of a regular dodecahedron after a one-tenth counterclockwise turn.[150] Recall that a regular dodecahedron is the twelve-sided polyhedron we get by assembling twelve regular pentagons of the same size to make a closed surface bounding a solid. It is the fifth and last Platonic solid. The Pythagoreans would have been delighted.

Fifty-three pages later, Poincaré concluded the fifth complement:

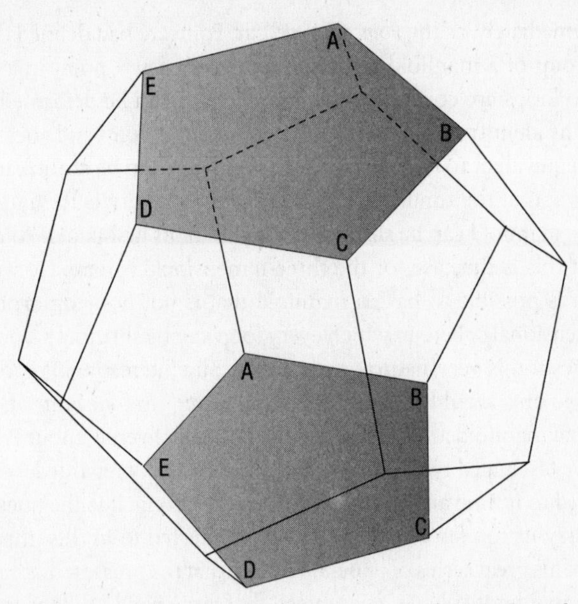

FIGURE 48. The Poincaré dodecahedral space. Connect the opposite faces after a one-tenth counterclockwise turn.

There are therefore two loops on the manifold which are not equivalent to a point; so the manifold cannot be homeomorphic to a sphere.

In other words . . . [He restates, explaining that the fundamental group is not the identity].

There remains one question to handle: Is it possible that the fundamental group of a manifold could be the identity, but that the manifold might not be homeomorphic to the three-dimensional sphere? [This is the Poincaré conjecture.]

In other words . . . [He carefully explains, in terms of the example he has constructed, what would need to happen to get such a manifold].

But this question would carry us too far away.[151]

The "one question to handle"—the question "Is it possible that the fundamental group of a manifold could be the identity, but that the manifold might not be homeomorphic to the three-dimensional sphere?"—became known, almost immediately, as the *Poincaré conjecture*. Poincaré had defined the fundamental group of a manifold to be the set of loops at a point in a manifold where two loops are considered the same if they can be deformed into one another. The identity is the loop that stays at a single point and goes nowhere. A loop is equivalent to the identity if and only if it can be shrunk to a point. Thus, saying that the fundamental group is the identity is to say that every loop in the manifold can be shrunk to a point. In his first paper, Poincaré had noted that this is the case for the three-dimensional sphere. He was asking whether it is possible to have a manifold that is not homeomorphic to the three-dimensional sphere in which every loop can be shrunk to a point.

The question is very natural and intrinsically interesting. It is one of the first things one would ask after encountering the notion of a three-dimensional manifold and the fundamental group. Even without Poincaré, it would have fascinated researchers. But Poincaré clearly regarded it as central. He returned to it time and again. He first got it wrong. It is the question that "would carry us too far away," the question referred to in this, the very last sentence of his great series of papers. The fact that the greatest mathematician then alive could not solve it, guaranteed that fame would follow any mathematician who did.

11

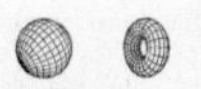

The Great Savants

As the new century began, Paris remained the social and cultural capital of the world. It also had the largest population of mathematicians. Most visible among them was Henri Poincaré, then the most famous mathematician of his time. From his Olympian vantage point at the top of both the French and world scientific pyramid, Poincaré was in touch with all the best ideas and had a hand in the era's greatest scientific achievements. His work ranged over almost the whole of mathematics and over large areas of physics. Between 1901 and 1912, Poincaré was nominated for the Nobel Prize no fewer than forty-nine times, more than any scientist before or since.[152]

Poincaré systematically refused any political engagement or duty, but seemingly never declined any service in the administration or service of science. His modesty and his gentle humor were a refreshing contrast to the edgier personality traits of some of his less accomplished colleagues. He was a member of the committee for the International Exposition of Paris (the one that gave us the Eiffel Tower) and he had become the head of the critical Bureau of Longitude. He was consulted increasingly about every major science policy decision. The Dreyfus affair was rocking France of the time, and Poincaré was appointed to a commission to investigate the scientific validity of the evidence that had been presented.

By then, Poincaré's fame had spread to the general French public. *Science and Hypothesis,* his first book written for a general educated readership, would sell more than sixteen thousand copies in France in the first ten years after its publication in 1902. Its success may have lead to the clever jape familiar to all French schoolchildren, *"Quest-ce un cercle? Ce n'est point carré"* (What's a circle? It's not a square), as *point carré* is pronounced the same as "Poincaré". The book would ultimately be translated into twenty-three languages. A second book for a general audience appeared three years later, and a third in 1908.[153] These books, in Poincaré's lovely prose style, resulted in the author's election to the French Academy in the chair vacated by Sully Prudhomme. Then and now, election to one of the forty positions in that academy is perhaps the highest accolade that can be awarded a French intellectual. Election to both the Academy of Sciences and the French Academy is exceedingly rare.[154]

Critics carped that Poincaré's work contained sloppy errors and that he did not care. Nothing was further from the truth. Far from having a thick skin, Poincaré would return to topics when he or others found errors in his results. He took pride in his work and would not publish until he had settled an issue to his own satisfaction. While it is true that he seldom revised what he had written, it was not from lack of care, but out of impatience. With so many ideas, with so little time, revision was a luxury that he could not afford.

The wonder is that Poincaré continued to publish so much, well after he had any need to establish his reputation. Virtually everything he published received heavy scrutiny. Mathematicians hung on his every word. Wherever he took interest in the work of another mathematician, it would boost that person's career. Conversely, to find or fix a gap in Poincaré's reasoning would also further one's career. He may have been revered, but he was also a target. The pressure must have been tremendous, and one can only admire the courage that it took for him to continue to publish.

His topological papers are a case in point. They have their origin in a mistake that Poincaré had made in his submission for the problem of stability of the universe. The Danish mathematician, Poul Heegaard, made his name by providing a counterexample to the first version of the Poincaré duality theorem. Poincaré's desire to repair his theorem occasioned the first complement. The fifth complement, and with it the Poincaré conjecture, arose because the Frenchman had himself discovered an error in his own work. No matter how beautiful his dodecahedral space, in a world where his every word was examined, such highly public errors must have been very painful to Poincaré. Whether he ever reflected on them in print, there can be little doubt that they informed his writing. In *The Value of Science* he wrote movingly of staying the course.

> The search for truth should be the goal of our activities, it is the sole end worthy of them.... But sometimes truth frightens us.... We also know how cruel the truth often is, and we wonder whether illusion is not more consoling, yea, even more bracing, for illusion it is which gives confidence.... This is why many of us fear truth; we consider it a cause of weakness. Yet truth should not be feared, for it alone is beautiful.... When I speak here of truth, assuredly I refer first to scientific truth; but I also mean moral truth, of which what we call justice is only one aspect. It may seem that I am misusing words, that I combine thus under the same name two things having nothing in common; that scientific truth, which is demonstrated, can in no way be likened to moral truth, which is felt. And yet I cannot separate them, and whosoever loves the one cannot help loving the other. To find the one, as well as to find the other, it is necessary to free the soul completely from prejudice and from passion; it is necessary to attain absolute sincerity. These two sorts of truth when discovered give the same joy; each when perceived beams with the same splendour, so that we must see it or close our eyes; they are never fixed; when we think to have reached them, we find we still have to advance, and he who pursues them is condemned never to know repose. It must be added that those who fear the one will also fear the other; for they are the ones who in everything are concerned above all with consequences. In a word, I liken the two truths, because the same reasons make us love them and because the same reasons make us fear them.[155]

TOPOLOGY AND THE POINCARÉ CONJECTURE

It has been said that Poincaré did not invent topology, but that he gave it wings. This is surely true, and verges on understatement. His six great topological papers created, almost out of nothing, the field of algebraic topology. The new discipline would drive some of twentieth-century mathematics greatest successes. Poincaré wrote to be understood, using plenty of examples, in a style that seems leisurely by today's standards. However, for mathematicians of the time, the sheer number of genuinely new ideas made reading his topological papers like drinking from a fire hose. His intuition let him range far from known areas with secure foundations. Every result, every page, drew one in and beckoned the unwary into dangerous intellectual shoals that would require the determined efforts of scores of mathematicians to explore. The fields of general and combinatorial topology grew up partly as an attempt to allow others to navigate regions over which Poincaré had first ranged.

The last statement of the fifth and last complement, the Poincaré conjecture, exerted a sirenlike influence on mathematicians. Here was the simplest question that arose in thinking about the shape of the universe. For those who ventured into the topological papers, the question became an obsession. The first victim, other than Poincaré himself, was the brilliant Max Dehn.

Dehn had made a name for himself as a student at Göttingen by solving the third in the famous list of twenty-three problems that Hilbert had proposed in 1900.[156] Dehn showed that one could not cut a tetrahedron up into a finite number of pieces with planar cuts and paste the pieces together to form a cube. This was a problem that had been unsolved since Euclid's time, and both Gauss and Hilbert had worked on it unsuccessfully. One consequence of Dehn's work was that one could not define the volume of a polyhedron without infinite constructions.[157]

At Göttingen, Dehn acquired a taste for rigorously formulating definitions and axiomatizing reasoning. Heegaard's counterexample to Poincaré's first formulation of the duality theorem pushed Poincaré in the direction of more combinatorial definitions of manifolds and homology groups which, in turn, lent themselves well to axiomatic formulations. Dehn and Heegaard had met at the International Congress of Mathematicians in Heidelberg in 1904 and began to work together.[158] They jointly wrote an article that carefully, and very abstractly, laid out the foundations of combinatorial topology for the *Encyclopedia of Mathematical Knowledge* project that Klein was overseeing.[159] It appeared in 1907 and gave the first absolutely rigorous classification of surfaces. It also contained an erroneous account of Poincaré's construction of his dodecahedral space: The authors had made an error that Poincaré had avoided. The example that they substituted for Poincaré's was in fact the three-sphere, not the dodecahedral space. Their error underscored the subtlety of Poincaré's construction. He must have tried many different possibilities to find his example.

Dehn was very impressed with Poincaré's 1904 paper and would return to it throughout his life. By 1908, Dehn thought he had succeeded in proving the Poincaré conjecture. He actually submitted his proof to *Mathematische Annalen* and wrote Hilbert urging him to speed up publication in case someone got there first. Like who? "Poincaré, for example," he wrote. However, after conversation with topologist Heinrich Tietze at the 1908 International Congress of Mathematicians in Rome, Dehn realized that his reasoning was erroneous and he withdrew the paper.[160]

Poincaré's dodecahedral space was the first example of a three-manifold whose homology groups are the same as those of the three-sphere, but which is not homeomorphic to the three-sphere. At the time, the simple description of the example in terms of identifying opposite faces of a dodecahedron was not available, and the space seemed totally mysterious. Such manifolds are called homology spheres and seemed to have come out of nowhere. Were there more of them? And, if so, was there a way of finding them?

In order to produce examples, Dehn invented a precursor of what we now call *Dehn surgery*. To describe it, recall first that a three-sphere can be thought of as two solid balls stuck together by identifying points on the spheres that bound the balls. You can see this by taking a solid ball in the three-sphere. Since the three-sphere can be thought of as the result of attaching two solid balls to one another by gluing them along their boundary (see chapter 4), the region outside the ball is also a solid ball. Likewise, if we take a solid (unknotted) torus in the three-sphere, then the region outside the torus is also a solid torus. Dehn surgery is the process of removing a solid torus from the three-sphere and stitching it back in differently. To do this we just have to map the torus bounding the solid torus that we took out to the (same) torus bounding what is left. There are lots of ways of doing this that are fundamentally different from each other. For example, we can cut the torus along a meridian, give it a full twist (or several) and then reconnect it with the original torus. Think of a misanthropic surgeon opening someone up, cutting through the intestine, giving it a full twist, stitching it back together, and closing the incision. Just as one might find oneself experiencing some difficulties after such a procedure, what results after Dehn surgery may not be homeomorphic to the three-sphere.

In 1910, Dehn and Heegaard published a famous paper that used Dehn surgery to produce an infinite series of three-manifolds that were homology spheres.[161] They close the paper by outlining an argument that they hope will lead to a proof of the Poincaré conjecture, while at the same time pointing out a critical gap that prevents the argument from going through. Both authors clearly believed the Poincaré conjecture was true. This paper made it clear to the rest of the mathematical community that, true or not, the conjecture was difficult.

The 1910 paper is interesting for a number of other reasons. It showed that there was a connection between homology spheres and non-Euclidean geometry.[162] It also investigated some connections between the theory of knots and three-manifolds. One of its most striking results relied on a notorious result, now called Dehn's lemma, that Dehn thought he had proved. However, the proof was subsequently found to contain a gap and Dehn's lemma was only proved in 1957.[163]

Tietze, the canny mathematician who had prevented Dehn from publishing his erroneous proof of the Poincaré conjecture, was one of the greats of the young field of topology. Austrian by birth, he acquired his interest in topology from the Austrian function theorist, Wilhelm Wirtinger. Wirtinger, who had been heavily influenced by Klein and was broadly interested in mathematics, had started using topological methods to investigate functions of two complex variables that were defined implicitly by a polynomial in three complex variables. Tietze's *Habilitationschrift* provided a lucid exposition and a coherent combinatorial approach to three-manifolds. Tietze also pointed out some of the basic questions that Poincaré's work left unresolved, and introduced distinctions critical for the development of the emerging field.[164]

RELATIVITY

Although twentieth-century topology began with Poincaré, and although he himself signaled its importance, it was a tiny part of his total work. Poincaré had a hand in some of the era's other great scientific achievements, and his analysis of his own scientific work, written in 1901, devotes only three of ninety-nine pages to topology. Similarly, topology occupies less than one of the seventy-four pages in the eulogy of Poincaré written by his dean, Gaston Darboux. Only two pages of Jacques Hadamard's eighty-five-page account of Poincaré's mathematical work deals with topology.[165] And fewer than ten of Poincaré's more than five hundred papers deal with topology.

Poincaré was a member of the Bureau of Longitude and spearheaded the bureau's quest to deliver synchronized time to the world. He oversaw the 1897 report on the decimalization of time, and was the liaison between the Academy of Sciences and the complicated longitude mission to Quito, Ecuador. He was one of the visionaries who used the Eiffel Tower to send out synchronized time signals, the precursor of today's GPS system, an innovative use of Paris's great iron mast that allowed uniform clock settings and the determination of longitude.

Partly as a result of this service, partly as a result of his interest in mathematical physics and celestial mechanics, Poincaré had thought deeply about the nature of time. In 1898, he wrote a paper asking whether one second today is equal to one second tomorrow, and whether it is meaningful to say that two events in different places happen at the same time. He was heavily involved in working out the consequences of experiments which seemed to suggest that distances contracted in the direction of motion. In 1905, a then-obscure patent clerk, Albert Einstein, burst onto the scientific scene with four major papers, all recognized today as classics. By 1909, Einstein would be recognized as a major thinker, on the order of a Poincaré. The relationship between the two men was complex: They met only once, at a conference in 1911 in Solvay, Belgium. Poincaré had a very high opinion of Einstein; Einstein saw Poincaré as one of the reactionary old guard who still clung to useless notions like the ether.[166]

Some credit Poincaré with the independent discovery of special relativity, citing his magnificent paper on the dynamics of the electron.[167] The interplay between the era's philosophical presuppositions, its pure science (especially physics), and its technological needs (the French had overseas possessions to administer; the Swiss, complex train schedules to coordinate), and the backgrounds and temperaments of Poincaré and Einstein has been nicely described in a recent book by the historian of science Peter Galison:

> Did Einstein really discover relativity? Did Poincaré already have it? These old questions have grown as tedious as they are fruitless. . . . Here were two great modernisms of physics, two ferociously ambitious attempts to grasp the world in its totality. . . . One [Poincaré's] was constructive, building up to a complexity that would capture the structural relations of the world. The other [Einstein's] was more critical, more willing to set aside complexity in order to grasp, austerely, those principles that reflected the governing natural order. . . . The Bureau of Longitude that Poincaré had helped supervise stood as one of the great time centers of the world for the construction of maps. And the Swiss Patent Office, where Einstein had stood guard as a patent sentinel, was the great Swiss inspection point for the country's technologies concocted to synchronize time in railways and cities.[168]

Like topology, the theory of relativity had begun to make large inroads on the scientific consciousness of the times. The key principle that the laws of physics must look the same to observers who were moving at constant velocity relative to one another had enormous consequences, among them that matter and energy were two manifestations of the same phenomenon, that time and space were thoroughly interrelated, dilating at higher velocities. Poincaré was probably the first to put space and time together as a mathematical object that came to be called *space-time*. Hermann Minkowski, of Göttingen, showed that work of Lorentz, Poincaré, and Einstein could best be understood in terms of a new non-Euclidean geometry on space-time. Einstein was initially skeptical. By 1912, however, he became more receptive as he sought to generalize special relativity to the context in which observers were in any sort of motion, and in particular accelerating, relative to one another. Ultimately, he would discover that one way to frame the general laws of physics was in terms of Riemannian geometry on space-time.

GERMANY AND GÖTTINGEN

After his breakdown in 1882 partly as a result of the overwork brought on by his rivalry with Poincaré, Felix Klein went on to build Leipzig into a major research center, even while plagued by depression. His worldwide reputation continued to grow. An offer to occupy the recently vacated chair in mathematics at the Johns Hopkins University, the first institution of graduate research in the United States, revived his spirits. He would almost certainly have gone had the president of the university met the $6,000 salary that the previous chair holder had been paid. Instead, the president remained firm on an offer of $5,000, and Klein declined it. It is tempting, though futile, to imagine how the mathematical profession in the United States would have developed had Klein moved to Johns Hopkins. Rarely has a whole generation's educational opportunity been lost for want of $1,000.

Partly in response to the offer from Johns Hopkins, Klein received an attractive offer from Göttingen in 1886, during a time when mathematics enrollments within Germany began to collapse. He was a gifted, intense, demanding teacher with an almost Faustian magnetism. Admission to his upper-level seminars was strictly limited. He did not suffer fools gladly, and only the ablest and those willing to work hard were admitted. His lectures were carefully prepared, and had a genuine sense of narrative. He would examine one telling example, working out just enough details to show the main lines of a representative argument. He would then use that example as a way to survey a whole panorama. He loved broad themes and stressed how different concepts related to one another in sometimes subtle ways. Technicalities bored him. His students took turns carefully writing up each lecture, filling in details (some of which were highly vexing and contained major mathematical problems that were not so easily resolvable).

Klein's broad scholarly view of mathematics contrasted with the intensely narrow research focus of some of his colleagues and drew students from outside Germany, especially the United States. He played a decisive role in the establishment of the American mathematical community by educating the mathematicians who would, in later years, help establish what would become major research institutions.[169] He was also very encouraging to women in mathematics and had several female doctoral students.

As gifted a teacher (and a researcher) as Klein was, his greatest influence would be as an administrator. He had a nose for mathematical talent and was unafraid of hiring individuals more talented than himself. His teaching and editorship of the influential journal *Mathematische Annalen* brought him into contact with the ablest young mathematicians of the time. He was intellectually generous, stayed in contact, and took an interest in young people's work, helping them secure positions. What was in effect a farm system, together with his willingness to take risks, his talent for intrigue and his friendship with Althoff, the Prussian minister of higher education, allowed him to make a series of brilliant hires.

At Göttingen, Klein's gift for creating scholarly community gave rise to a mathematical Camelot. The turning point was his hiring of David Hilbert, his gifted former student from Leipzig. Klein had stayed in contact with Hilbert after leaving for Göttingen, and encouraged him to visit Paris, where he met many French mathematicians, including Poincaré. On returning to Germany, Hilbert became interested in one of the most significant mathematical problems of the day, the *Gordan problem*, which abstracts the experience by which we recognize such objects as people, trees, and places even after changes are wrought by time and point of view. In mathematical terms, one thinks of a group of transformations acting on a set of mathematical objects, usually described by equations, and asks whether there are some objects or quantities, typically described by algebraic expressions, that remain invariant under the action of a group of transformations. Because there were so many sets of equations that one could fruitfully focus on, and so many groups of transformations that were geometrically significant and computationally accessible, questions of computing invariants for given sets of equations and a specific group were a minor industry, providing hundreds of subtle, but relatively accessible, problems for mathematicians and their students.

The study of invariants was popular in England as well as Germany, and the first American graduate students at Johns Hopkins, under the influence of British expatriate J. J. Sylvester, had worked out invariants for large numbers of mathematical objects. The acknowledged king of this whole enterprise was Paul Gordan, a friend of Klein's and a professor at the University of Erlangen. He had the best general result on invariants. As a result of these efforts, the general structure of such invariants was known for a relatively wide set of equations and a number of groups.[170] The *Gordan problem* asked if, for any set of equations and any group, it were possible to find a finite set of invariant expressions in terms of which all others could be worked out. By brute force computation, Gordan showed that this was indeed the case for all equations in two variables and a broad class of groups. This was considered a stunning achievement.

If the answer to the Gordan problem were affirmative, everyone tacitly assumed that the solution would involve explicitly working out invariants for representative sets of equations and groups, patiently showing that the list of invariants was complete in each case, and then showing somehow that the result for the representative set implied it for every group and every set of equations. The task was enormous, and the work would extend for generations. In a four-page paper published after a year spent wrestling with the problem, Hilbert completely solved the problem by showing that assuming there was no such finite basis led to a contradiction. Therefore, it was logically necessary that such a basis exists. The first reaction was stunned disbelief. "*Unheimlich,*" muttered Lindemann, one of Hilbert's older colleagues. "*Das ist nicht Mathematik. Das ist Theologie,*" pronounced Gordan.[171] Klein was enchanted.

Despite the relentless ferocity of his reasoning, Hilbert was too good a mathematician not to be interested in how to work out such sets of invariants, and he made substantial progress, again using methods that seemed totally foreign to researchers in the field.[172] He wrote everything up, and at Klein's invitation, submitted it to *Mathematische Annalen.* Gordan, the expert appointed to referee the paper, complained that Hilbert's standard of truth was not that he had shown something beyond any possible doubt, but rather that no one could contradict him. On learning about Gordan's report, Hilbert wrote to Klein saying that he was not prepared to alter anything, and this was to be his last word on the subject unless someone could show that his reasoning was erroneous. The youthful determination and firmness must surely have reminded Klein of his dust-up with Poincaré. Hilbert was reasonable, but there was a point beyond which he could not be pushed, and he put Klein in a tight spot. Klein sided with Hilbert, finding his thought "wholly simple, and therefore logically compelling." He also decided that Hilbert had to come to Göttingen.

In 1895, Klein finally succeeded in hiring Hilbert, and Göttingen became the strongest mathematical center in Germany. Students quickly discovered Hilbert. He was interested in details and would occasionally get stuck, but watching him work his way through difficulties was instructive. His lectures were elegant, but not as polished as Klein's. Where Klein would range broadly, Hilbert would focus. He was a minimalist and sought the shortest way into a subject. One could learn from Hilbert's lectures.

With the hiring of Hilbert, Göttingen acquired a personality and a mathematician of sufficient heft to outweigh Klein. While Klein loved the grand statement and happily engaged in political intrigue and the weighing of consequences, Hilbert distrusted sweeping statements and was utterly direct. Klein was an impresario, and Hilbert a stunningly capable, no-nonsense mathematician. Together, the two were formidable.

Mathematicians flocked to Göttingen and one spectacular appointment followed another. The University of Berlin, nominally the most prestigious in Germany, attempted to hire away Hilbert. He stayed.

Hilbert's research interests changed over the years. His modus operandi was to completely change fields every decade or so, and to make fundamental advances in each field in which he worked. He began in invariant theory, switched to algebraic number theory, and near the turn of the last century became very interested in the foundations of geometry. Hilbert completely rewrote Euclid's *Elements*, putting the entire development on a completely rigorous basis. He pushed to formulate axioms that were so crystal clear that they did not admit any ambiguity. He argued that the axioms had to be so complete that if one were to everywhere replace Euclid's terms "points," "lines," and "planes" by "beer," "table-legs," and "chairs," then the whole theory would go through. One could not rely on intuition to fill in the gaps.

FIGURE 49. David Hilbert

Hilbert's little book on the foundations of geometry became a best-seller.[173] It brought new life to a very old subject, and he pushed far beyond Euclid. He not only introduced new axioms to capture notions like betweenness and order, whose properties Euclid had tacitly invoked, but he varied the axioms obtaining different geometries.

Poincaré was delighted, writing a careful approving review showing that he had clearly engaged with the material and thought through the implications. He worried a bit about Hilbert's fascination with logic at the expense of the geometry: "The logical point of view alone appears to interest him." He concluded: "His work is then incomplete; but this is not a criticism which I make against him. Incomplete one must indeed resign one's self to be. It is enough that he has made the philosophy of mathematics take a long step in advance, comparable to those which were due to Lobachevsky, to Riemann, to Helmholtz, and to Lie."[174] Poincaré was not given to overstatement. This is a rave.

Hilbert was able to show that Euclidean geometry was consistent provided that arithmetic was consistent. Hilbert, like Klein, was interested in almost all areas of mathematics. He turned to mathematical logic, making fundamental contributions that would initiate some of the twentieth-century's greatest achievements, but that are outside the scope of this book. In 1900, he gave an address at the International Congress in Paris in which he listed twenty-three problems that he thought would provide an agenda for mathematicians for the next century. The list was enormously influential. In subsequent years, Hilbert would become increasingly interested in mathematical physics, and Göttingen became critical in the development of the new theories of relativity and quantum mechanics. By 1910, Göttingen could boast hundreds of mathematics students from around the world, and a list of postdoctoral fellows (privatdozents) and instructors that read like a who's who of mathematical physics. Women, Jews, any nationality, all were welcome.[175] Hilbert had even invited Poincaré, who gave five lectures at Göttingen, four in German.

POINCARÉ'S DEATH

Poincaré fell seriously ill at the International Congress of Mathematicians in Rome in April 1908. He spent much of his time confined to bed as a result of an enlarged prostate, and was unable to deliver his lecture, which was read by Darboux. He required surgery and his wife, Louise, traveled to Rome to accompany him back to Paris. Although he recovered somewhat and resumed working, he was not really well.

He seems to have had some presentiment of his death. In December 1911, he sent a letter to the editor of the journal that had published his fifth complement and the paper on the motion of an electron.

> My Dear Friend, I have already spoken to you on your last visit of a paper that I have held onto for the last two years. I have not made any more progress on it and I had provisionally decided to let it rest for a while to ripen. This would have been fine if I could be sure of being able to return to it someday. But at my age, this is not the case.

The editor urged publication, saying that Poincaré could preface the paper with an explanation. The paper appeared in 1912 and opened with an apology from Poincaré: "Never before have I publicly presented work that is so unfinished."[176] He went on to say that a number of problems in dynamics relating to the existence of periodic solutions of the three-body problem depend on a simple geometric result that he had become more and more convinced is true, but that he had been unable to prove. He hinted that he did not have a lot of time left and hoped that other mathematicians would have more success than he. The episode, incidentally, confirms the care that Poincaré took with his publications. He may not have spent much time revising, but he was scrupulous about his results. The paper was Poincaré's last geometric paper, and is the beginning of yet another new field, *symplectic topology*, which studies manifolds with an additional structure that allows one to define areas of surfaces in the manifold (although not necessarily lengths of curves).[177]

A second surgery in July 1912 seemed to have been fully successful. Alas, an embolism finished him while he was dressing in Paris on July 17, 1912. His unexpected death at the height of his powers shocked the world. Tributes poured into Paris. "Henri Poincaré was truly the living brain of the rational sciences," wailed an obituary in the *Le Temps*.[178] His funeral in Paris was attended by heads of state and representatives of every major university. France's ablest mathematician, Jacques Hadamard, just five years younger than Poincaré, was drafted to prepare an account of Poincaré's work. Göttingen prepared a retrospective. Gaston Darboux delivered a lengthy eulogy documenting his contributions in many fields.

Poincaré was fully immersed in his times, and his work in the service of science and France characterized the best of his era. His premature death seemed to have some deeper significance that his eulogists dimly grasped. What is more apparent a hundred years later was that Poincaré had planted the seeds of an entirely different mathematics. And it was not just mathematics. His 1904 address on the future of mathematical physics was eerily prophetic.[179] Unlike Riemann, he died famous with many admirers. But like Riemann, he had no students, no school. Large portions of Poincaré's mathematical work were not understood at the time. The man was dead, but his ideas and his conjecture lived on. Although, events on the larger stage intervened to forestall the immediate continuation of his work by others, the ideas and problems that Poincaré was addressing would come to be fully appreciated decades later. If Hilbert shaped the agenda, Poincaré shaped the form of twentieth-century mathematics.

12

The Conjecture Takes Hold

At the time, the first few years of the twentieth century seemed to point to a new era, one in which internationalism and shared values guaranteed increasing prosperity, and a global world order with Europe firmly at the top. World fairs, numbering more than a hundred in the three decades beginning in 1880, sprang up to celebrate learning, culture, and technological progress.[180] Mathematics, like the sciences, had become thoroughly professionalized and echoed the exuberance of the times. The great centers remained the German universities and Paris, but the creation of national mathematical societies (Moscow in 1864, London 1865, France 1872, Tokyo 1877, Palermo 1884, New York 1888, Germany 1890) reflected the presence of serious mathematics elsewhere. Poincaré and his work exemplified the way in which the strong commitment to internationalism coexisted with unwavering patriotism. He was French to the core, but his outlook was international. The problems on which he worked drew inspiration from, and contributed to, a mathematical community that likewise saw itself as determinedly international.

Palermo, on the west of Sicily's north coast, was typical of the new age. It housed a wealthy middle class and a burgeoning cultural scene.[181] The numerous Greek temples in Sicily (there are more there than in Greece) suggested nearly indestructible cultural and intellectual roots and linked with a hallowed past. The mathematical society in Palermo grew, from its founding in 1884, to be the largest mathematical society in the world. Its journal, the *Rendiconti del Circolo matematico di Palermo*, was the international journal with the widest circulation.[182] The first and fifth complements, and the Poincaré conjecture in particular, appeared in it. It seems fitting somehow that what we now call Poincaré's three-dimensional dodecahedral space made its first public appearance near the place where the dodecahedron had been discovered millennia before: Palermo is a mere two hundred miles from Croton, where Pythagoras and his circle settled.

But Poincaré had also embodied the age's contradictions. An internationalist who had believed passionately in universalist, rationalist Enlightenment values, he was also certain that France and the Third Republic had a lock on these values. Iconically French, he had never outgrown the sense of shame and national catastrophe that stemmed from 1870.[183] In the end, the rot of nationalism would overwhelm the era's breezy internationalism. Less than two years after Poincaré's death, Gavrilo Princip, a member of a Serbian secret society, shot the Austrian archduke Franz Ferdinand on June 28, 1914. Interlocking treaties drew nation after nation into the tragedy that was the Great War. Austria mobilized against Serbia. Russia mobilized in support of Serbia against Austria. Germany mobilized in support of Austria against Russia. France mobilized against Germany in support of her ally Russia. Germany preemptively attacked France and, by the end of August, over a dozen countries had declared war. At the time, the conflicts came "not as a superlative tragedy, but as an interruption of the most exasperating kind. . . ."[184]

The Great War was far more than an interruption. Among its many devastations, it polarized mathematicians and ruptured feelings of international accord. National pride and rivalry hardened into hatred. Klein signed a declaration by some of Germany's famous scientists and artists, which pronounced their support of the kaiser and averred that a set of statements made by Germany's enemies about Germany were false. He was promptly expelled from the Academy of Science in Paris, never to be forgiven for his signature on the declaration. France's most influential mathematicians became rabidly anti-German.

The war destroyed the old order and fatally weakened Europe. A whole generation of promising European mathematicians disappeared in the carnage. Economic conditions destroyed what bombs and gas did not. Palermo's wealth dispersed, and the *circolo matematico* spiraled downward. Hyperinflation in Germany, partly the result of reparations payments, undermined civil society. Revolution in Russia created opportunities for sociopaths who would inflict a series of tragedies on the Russian people. The economic dislocations and the desire for order created conditions ripe for extremism, totalitarianism and another war.

In retrospect, the optimism of the first years of the twentieth century seems unbearably naïve, and the internationalism impossibly fragile. The energy of the times now appears less that of health and youth, than the doomed frenzy of the last stages of consumption. The cultural exhibits of the world fairs celebrating the glory of the human spirit sat side by side with massive displays of armaments.

GENERAL RELATIVITY

Not all was lost. The life of the mind, of course, can never be stilled, no matter how deafening the cannonade. Just prior to the war, Einstein had moved from a full professorship in Zurich to a pure research position in Berlin.[185] In looking to extend relativity to contexts in which reference frames were accelerating relative to one another, he needed a mathematical language that allowed very general sets of transformations. With the help of Marcel Grossman, a mathematician and friend with whom he had worked in Zurich, he found that Riemann's work perfectly suited the physical principles that he was uncovering: It allowed for variations in curvature and geometry from point to point, and the machinery developed by the Italian geometers Gregorio Ricci-Curbastro and his student Tullio Levi-Civita allowed Einstein to reframe the differential equations that expressed the classical laws of physics. In Einstein's general theory, the forces due to acceleration are indistinguishable from those of gravity, hence equal. If one views oneself as accelerating relative to some fixed point of reference, one will express the force one experiences as coming from the acceleration. If instead one views oneself as stationary relative to a fixed point, one will express the force as gravity. Acceleration is always relative to something else, and the forces due to acceleration and gravity are equivalent. Einstein expressed gravity as the curvature of space-time (and according to Riemann, curvature is a tensor that expresses the deviation from 180 degrees of the angle sums of geodesic triangles at any possible orientation to an observer). The Einstein equation describes how the curvature tensor varies in the presence of matter. Matter curves space-time. By 1915, as a young generation lay dying in the trenches, he had the general theory that would shape the worldview of generations to come. "To my great joy," he wrote, "I completely succeeded in convincing Hilbert and Klein."[186]

Einstein's work became the rage. He would become a world celebrity in 1919 when a British scientific expedition observing a solar eclipse measured that light passing the sun "bent" precisely in accordance with his equations. The light, of course, did not bend—light followed a geodesic in space-time. What happened was that the massive sun curved space-time around it, and the path that light took appeared to bend as a result.

Einstein's work enormously stimulated the development of Riemannian geometry. At that time, it did not directly affect the development of topology. Einstein was certainly aware of the existence of different three-dimensional manifolds, and the possibility that space and space-time could have different topologies. But the equations of general relativity were differential equations and applied to small regions of space-time. Topology applied to the large-scale structure of space (and of space-time). No one dreamed that there might be a connection between the Poincaré conjecture and general relativity.

THE POINCARÉ CONJECTURE BETWEEN THE WARS

After Dehn and Tietze's work, there could be no doubt that the Poincaré conjecture was difficult, and that Poincaré's topological work was a significant source of interesting problems. Dehn and Tietze provided a possible starting point from which topologists could begin to approach Poincaré's work and the conjecture in particular.

The most decisive advances came from an American, James W. Alexander (1888–1971), who had gone to Europe for postdoctoral studies after receiving his degree from Princeton. Alexander had stayed in Paris during the war and translated Heegaard's thesis into French.[187] He joined the U.S. Army in 1917, rising to rank of captain, and retired in 1920 to join the faculty of Princeton University. In 1919, Alexander showed that two three-manifolds that had been found earlier by Tietze, and that had the same fundamental group and same homology groups were not homeomorphic.[188] Thus, the question of whether two three-manifolds with the same fundamental group necessarily were homeomorphic had a negative answer. The Poincaré conjecture is the special case of this question: when the fundamental groups have only a single element.

Alexander's result upped the stakes enormously on the Poincaré conjecture, and raised the distinct possibility that the conjecture was false.

Alexander went onto make an enormous number of critical discoveries.[189] At the International Congress of Mathematicians in 1932, he gave one of the main addresses and in it underscored the importance of the Poincaré conjecture.[190] Shortly thereafter, J. H. C. Whitehead (1904–60), the easygoing and well-loved British mathematician who trained at Princeton and subsequently began the strong tradition of topology at Oxford, announced a proof of a theorem that would prove the Poincaré conjecture.[191] The result got past the reviewer, and appeared in print. Shortly thereafter, Whitehead discovered a counterexample to his theorem. Whitehead's counterexample, like Poincaré's, taught us much, but said nothing about the conjecture. The score stood at: Poincaré conjecture 3, Mathematicians 0. Friends attribute the almost painful precision in Whitehead's subsequent papers to this embarrassment.

By 1936, the Poincaré conjecture was one of the best-known problems in mathematics, and topology had emerged from adolescence. Two influential texts, both the result of interesting collaborations, were published that made it possible to understand much of Poincaré's work by carefully filling in the basics.[192] Both books went further by significantly extending some of Poincaré's results and both mentioned prominently the Poincaré conjecture.

The first, *Lehrbuch der Topologie*, by Seifert and Threlfall, appeared in 1934. It developed the polyhedral approach, and was full of lovely examples and notes. The book originated when twenty-year-old Herbert Seifert enrolled in William Threlfall's course on topology in the Technische Hochschule in Dresden. Seifert's imagination was set on fire, and the two men became fast friends and collaborators. Seifert spent the 1928/29 term at Göttingen, where he met Heinz Hopf, who was a postdoctoral fellow, and Pavel Aleksandrov, who was visiting from Moscow and who would later found the great Moscow school of topology. Seifert returned to Dresden, moving into Threlfall's house, and the two men worked nonstop on three-dimensional manifolds.

According to his diary, Threlfall wanted the preface of the textbook that grew out of his original lecture notes to begin "This textbook arose from a course that one of us gave the other at the Technical University of Dresden, but soon the student contributed so many new ideas and so fundamentally altered the presentation, that the name of the original author might better have been omitted from the title page." The actual preface reads, "The original stimulus to the writing of the present textbook was a series of lectures which one of us (Threlfall) presented at the Technical University of Dresden." The book developed the principal concepts of topology through carefully worked out examples. It contains a careful discussion of Poincaré's dodecahedral space and calls attention to the Poincaré conjecture: "Whether the 3-sphere is characterized by its fundamental group is the content of the 'Poincaré conjecture' which remains unproven to this day." Lest the reader miss the point, they elaborate: "Aside from the 3-sphere, do there exist other 3-dimensional closed manifolds such that each path can be contracted to a point?"[193]

The second text, *Topologie*, was written by Aleksandrov and Hopf, and was in press during the international topology meeting in Moscow in 1935. This book was a bible of sorts, even though it was the only one of three projected volumes to have been completed. Aleksandrov began collaborating with Hopf at Göttingen in 1926. Hopf had earlier classified simply connected three-manifolds of constant curvature as part of his Ph.D. at the University of Berlin. The two spent the 1927–28 academic year at Princeton. Aleksandrov became a full professor at the University of Moscow in 1929. The same year, Hopf was offered an assistant professorship at Princeton, but elected to stay in Europe, and received a chair in Zurich in 1931. Clearly, a tentative internationalism was arising from the ashes of the interwar period. More important for our purposes, their little bible of a book stated the Poincaré conjecture right in its introduction.

THE EMERGENCE OF MATHEMATICS IN THE UNITED STATES

James W. Alexander, the American in Paris, was a student of the influential Oswald Veblen. That both mathematicians were American signaled that a profound shift had taken place in the United States. Prior to the dawn of the twentieth century, no American university could convincingly claim to be of world class, and no American could get an adequate mathematical education without going abroad.[194] If Riemann or if Poincaré had been educated in the United States, they could not have developed into the mathematicians that they subsequently became. This would change in the new century, and mathematics would never be the same.

Some of the distinctive features of American higher education, however, were in place by the beginning of the century. A high-level commission chaired by the president of Harvard in 1892 rejected the establishment of national standards for secondary schools, and by extension, admission standards for postsecondary education. By the end of the nineteenth century, the United States had an enormous number and variety of institutions of higher learning, both public and private. The thirty years after the Civil War had seen a huge growth of wealth, and the greatest boom in higher education in the nation's history. New state institutions were founded, along with women's colleges and colleges for African Americans. These institutions needed faculty. Moreover, Americans were ambitious and recognized the need for graduate education and research. Some of the older universities began experimenting with the establishment of German-style Ph.D. programs,[195] and new graduate institutions were founded. But the quality was lacking: Establishing truly successful graduate programs, in mathematics at least, had eluded Americans. Mathematicians are a bit like computer chips. To get great ones, one needs lots of good ones. And, in the case of mathematicians, it helps if the very good ones have a knack for recognizing and encouraging talent.

It was in Chicago that the right combination of money, talent, serendipity, and effort came together. The American Baptists and the citizenry of Chicago had decided to establish a university in Chicago, and Standard Oil founder and CEO John D. Rockefeller, a devoted Baptist, funded it. From the outset, the University of Chicago took as its mission the advancement of knowledge through research and graduate education, as well as the provision of undergraduate education. Rockefeller hired William Rainey Harper, a professor at Yale Divinity School, as president. Even in an age of great university presidents, Harper stood out as a visionary administrator with an eye for quality and the bottom line. He hired a talented young Midwesterner, E. H. Moore, away from Yale, and they worked together to establish a first-class mathematics program. Partly as a result of a raid of rival Clark University still remembered bitterly today, they acquired two German émigrés, both former students of Klein, for the mathematics faculty. Something magical came together, and Chicago emerged as an astonishingly productive mathematics department, both as measured by numbers of research papers, but more important for the United States, mathematicians. The only other rival, not nearly as effective except arguably in the field of analysis, was Harvard, where another two students of Klein held forth.[196] Chicago doctoral graduates went on to build up major mathematics departments throughout the country.[197]

Two of Moore's students, Oswald Veblen (1880–1960) and George D. Birkhoff (1884–1944), had been undergraduates at Harvard and had gone to the University of Chicago for their doctoral work. Veblen graduated in 1903, and Birkhoff in 1907. After two additional years as an instructor at Chicago, Veblen moved to Princeton in 1905 in one of the positions that then-Princeton, and future-U.S., president Woodrow Wilson had set up in an attempt to raise the research level of the university. By 1910, Wilson had become governor of the state of New Jersey; the head of the math department, Henry Fine, had become dean of faculty; and Veblen, a full professor. Birkhoff moved to Harvard in 1912 from an assistant professorship at Princeton.

Klein may have trained the senior mathematicians at Chicago and Harvard, but it was Poincaré who fired the imaginations of the young. Veblen's thesis had been on the axioms of geometry, but he moved over to topology and relativity in the years following his doctorate. His first topological work appeared in 1905 and mixes the axioms of geometry with plane topology.[198] Veblen introduced Poincaré's topological ideas to American mathematicians in his book of 1922. Birkhoff independently read Poincaré's work on dynamical systems and was much taken with the Frenchman's use of topological methods in understanding systems of differential equations. In 1913, he became famous for proving Poincaré's last geometric theorem, the conjectural theorem that Poincaré had been unable to prove and had reluctantly published the year he died (1912). In Göttingen, there was incredulity that an American had solved it.

Both Birkhoff and Veblen worked tirelessly on behalf of mathematics. Birkhoff became a vice president of the American Mathematical Society in 1919, and dean of arts and sciences at Harvard in 1936. His mathematical influence was enormous, and some of his students became mathematicians who will be remembered for centuries hence. It was Veblen, to a large extent, who made Princeton into one of the world's great universities for mathematics. The department concentrated on a few, very current fields: topology, differential geometry, mathematical physics, and mathematical logic. By the First World War, Princeton had begun to rival European institutions such as Göttingen and the two universities in Vienna in topology. By the late 1920s, there was no contest: Princeton was the leading institution in the world, hands down. Veblen understood the importance of raising money from alumni and from private sources, and was extremely effective on both counts.[199] He oversaw the construction of a new mathematics building that became a magnet for mathematicians worldwide. Veblen played a key role in the conception of the Institute for Advanced Study and in locating it in Princeton. He resigned in 1932 from the named research professorship that he had held at university, becoming the first professor at the Institute for Advanced Study.

Both Veblen and Birkhoff were internationally known, and cultivated contacts throughout Europe. However, in his efforts to build up the American mathematical community, Birkhoff tended toward a sort of protectionism, arguing in print that efforts to find refugee European mathematicians positions at institutions in the United States took scarce university jobs away from Americans.[200] Tragically and stupidly, he also sought to limit the number of Jews in academic positions at Harvard and elsewhere. Veblen, in contrast, created a much more diverse community. Princeton both cultivated the young, promoting from within, and took advantage of the Nazi situation to attract Albert Einstein, John von Neumann, and Hermann Weyl to faculty positions. By the mid-1930s, Princeton had five legendary topologists[201] and many promising young ones. The United States had arrived: Serious study of topology meant spending some time in Princeton.

In addition to Alexander, an early brilliant hire of Veblen's was Solomon Lefschetz (1884–1972), a Russian Jew who emigrated from Paris in 1905. He worked as an engineer for Westinghouse Electric Company, until he lost both hands in a boiler explosion and could no longer practice engineering. He went back to school, receiving his Ph.D. in mathematics from Clark University in 1911. He then went to Nebraska for two years, and to Kansas for eleven. Far from the mathematical mainstream, he worked in comparative isolation and pushed Poincaré's applications of topology to algebraic geometry far beyond anything Poincaré did. He moved to Princeton in 1924. Loud, excitable, with two metal hands sheathed in a black plastic, and badly dressed, he was everything that many of the more cultured Princetonians were not. Lefschetz would get carried away with enthusiasm in mathematical conversation, excitedly pursuing ideas to their end, oblivious of surroundings or social convention.[202] People would stay away from him at parties.

Lefschetz was four years older than Alexander, and the two fell out as time went on. Alexander was wealthy and socially adept. Lefschetz was neither. Alexander resented what he felt to be Lefschetz's appropriation of his ideas without proper attribution. Lefschetz was all drive, Alexander singularly lacking in ambition. Lefschetz never got over Veblen's choice of Alexander instead of him for one of the first professorships at the Institute for Advanced Study. This was in spite of Lefschetz's getting Veblen's chair, which had no teaching duties.

For all that, Lefschetz was magnificent. He welcomed all, his appetite for mathematics was enormous, and his standards were high. He made Princeton truly excellent. Lefschetz was also impossible. As chair, he would always have to bring a more diplomatic colleague on trips to the dean's office. Lefschetz would go in thinking he had an airtight case for more faculty lines. The dean would not see it that way. Lefschetz would get excited. Words would be exchanged. The more diplomatic colleague would be needed to pull Lefschetz out before relations between the administration and the mathematics department suffered irreversible damage.

THE RUSSIAN SCHOOL

Russia had a strong tradition of mathematical research dating back to the Saint Petersburg Academy, which was founded in 1725. The counterpart of Veblen was Nikolai Nikolaevich Luzin (1883–1950). He nurtured a generation of mathematicians through the war and the Russian Revolution, and his research group was known among his students as the Luzitania. It included several very fine topologists who would contribute to making the University of Moscow one of the strongest centers of mathematics in the world. His first student was Pavel Aleksandrov, whom we met earlier. Another was Andrei Kolmogorov (1903–87), one of the most famous mathematicians of all time. Already as an undergraduate, Kolmogorov started producing important results. He published eight papers in his senior year, and eighteen by the time he received his doctorate. Many are still regarded as classics. Alexandrov and Kolmogorov became very close. In 1935, they bought a small house in the village Komarovka outside Moscow, where they hosted many mathematicians. In 1938, Aleksandrov and Kolmogorov and a number of other mathematicians at the University of Moscow joined the Academy of Sciences (the Steklov Institute) while retaining their positions at the university.

In 1935, the first international congress devoted completely to topology was held at the University of Moscow. Eight Americans, mostly Princetonians, attended. Several major breakthroughs were announced. The most surprising concerned the discovery of a new set of algebraic structures associated with manifolds and other topological spaces. These structures, called *cohomology rings*, were a kind of mirror image of the homology groups that Poincaré had defined, but they carried two algebraic operations instead of the single one carried by the homology groups. These carried more delicate topological information, and were the biggest advance since Poincaré. They also clarified some mysterious remarks of Poincaré's.[203] Kolmogorov announced the discovery. In the next lecture, Alexander confessed that he, too, had obtained virtually identical results and was going to talk on the same thing. Both men had already submitted for publication.

Sadly, the year after the congress saw the intensification of Stalin's purges. Luzin came under attack through the newspaper *Pravda*. He was accused of anti-Soviet propaganda and of publishing his important papers abroad instead of in Russian journals. He narrowly survived. He had to resign from the University of Moscow, but kept his position at the Academy of Sciences. The effect of his persecution was that Soviet mathematicians stopped publishing in the West and began to publish exclusively in Russia in Russian language journals. This isolation disserved both Western and Soviet mathematicians for years to come.

GERMANY AFTER THE WAR

When the First World War broke out, Göttingen emptied as individual after individual was called up. After the war, under the leadership of Richard Courant, it began to regain some of its former luster. Germany had massive financial problems, but the war erased any doubts about the potential of science and technology for creating industrial capacity, and for designing and building weaponry. A new mathematics building was erected and became the focus of the lively mathematics scene.

With Poincaré dead, Hilbert was the leading mathematician in the world. He had refused to sign the declaration that Klein had signed, a position for which he had taken a fair amount of heat from the Göttingen students and townspeople. But that would be only a mild foretaste of what was to come. The economic malaise induced by reparations payments and the global recession had radicalized the German electorate. In the Reichstag elections of 1932, the National Socialist Party made great gains. President von Hindenburg appointed Adolf Hitler the chancellor of Germany. The horror began and the lights dimmed. The universities were ordered to remove every so-called full-blooded Jew who held any sort of teaching position. Richard Courant, Edmund Landau, Emmy Noether, Paul Bernays. Gone. More would follow, including those with a Jewish ancestor or spouse. Sitting next to the Nazis' newly appointed minister of education at a banquet, Hilbert was asked, "And how is mathematics in Göttingen now that it has been freed of Jewish influence?"

"Mathematics in Göttingen?" Hilbert replied. "There is really none any more."[204]

13

Higher Dimensions

The Second World War, coming barely twenty years after the First, devastated another generation in Europe. The United States had losses to be sure, but they were not on the same scale as those in Europe. It emerged from the war buoyed by optimism and idealism.

As in the First World War, many mathematicians contributed to the war effort and had turned to areas of mathematics that were more immediately applicable to warfare. They played a large role in the development of radar, of the atomic bomb and of nuclear power, of coding and decoding, of jet flight and aerodynamics. On the one hand, involvement in the war had stripped mathematicians of their innocence, and many would subsequently be troubled by the ethical implications of what they produced. On the other hand, no one could doubt the efficacy of mathematics and of science, and mathematicians' success had tremendously raised the stakes in research. The National Science Foundation was formed in 1952 with a mission to support basic science. Various other federal government agencies in the United States began to fund science and mathematics.

Returning soldiers flooded the nation's universities on the GI Bill, triggering the greatest boom in higher education since just after the Civil War. In typically American fashion, great vision competed with savage narrow-mindedness. The Marshall Plan underwrote the rebuilding of Europe, and a Supreme Court decision outlawed segregation in American public schools, creating the as-yet-unrealized promise of access to first-class elementary and secondary education for children of all races, of either sex, and of any economic background. The cold war set in, and the House Un-American Activities Committee threatened to extinguish freedom of thought.

FOUR AND MORE DIMENSIONS

There was not as much of a lull in pure mathematics, and topology in particular, as it appeared at first blush. There was activity if one looked carefully enough, but the war had disrupted communication. Individuals continued to think about mathematics, but it took time for things to go back to normal, and for lives and universities to be rebuilt.

The dam burst in the late fifties. By 1960, we were in the midst of the most productive and most explosive era in the development of mathematical thought in human history. Nothing rivaled it: Not the courts in Babylon, not the Greek schools in Athens and Alexandria, not the Renaissance or Enlightenment Europe or nineteenth-century Germany. The outbreak occurred around the globe, in almost every field of mathematics. The core areas of geometry, topology, algebra, and analysis expanded enormously, and new disciplines on their edges and within subareas of each blossomed into fields with their own powerful methods and spectacular results. Advance after advance in computing, information sciences, and applied mathematics both fed, and were fed by, the staggering growth of mathematical knowledge.

From the point of view of topology and the Poincaré conjecture, the event that seemed to crystallize the change was an unexpected discovery by John Milnor in 1956. Milnor had received his bachelor's degree from Princeton in 1951 and his doctorate in 1954, and had stayed on as a faculty member. As an undergraduate he had solved a long-standing problem about mathematical *knots* (closed curves in a three-dimensional space).[205] Legend has it that he had mistakenly thought that it was a homework problem. A few years later in 1956, at not quite twenty-five years of age, Milnor used some recent, still undigested, work by the French mathematician René Thom to show that there were fundamentally different ways of doing calculus on a seven-dimensional sphere. This caught the imagination of mathematicians everywhere and opened a whole new world.

Let me explain briefly. Just as there are Euclidean spaces of every dimension, there are spheres of every dimension. The usual two-dimensional sphere can be thought of as the set of points of fixed distance, say one, from the origin in three-space and the three-sphere as the set of points of distance one from the origin in four-dimensional space. (Lest four-dimensional space seem unreal, remember that a point of four-dimensional space is just given by specifying four numbers, and four-dimensional space is just the set of all four-tuples of real numbers.) Likewise, the seven-sphere is the set of points of distance one from the origin in eight-dimensional space (the space of all eight-tuples of real numbers) or any set homeomorphic to it. Just as in the case of two- and three-dimensional spheres, the seven-dimensional sphere exists independently of the Euclidean space in which it is embedded. In general, for any positive integer *n*, an *n*-dimensional sphere (or *n*-sphere) is any set homeomorphic to the set of points of fixed distance from the origin in the Euclidean space of one dimension higher.

Any time that one has a manifold, one has another class of mathematical objects, called *functions*, on the manifold. A function is any rule that assigns numbers to different points of the manifold: Different assignments give different functions. *Calculus* is the study of rates of change, called *derivatives*, of functions. It studies how functions can change, and how one can determine a function if you just know its rates of change. There is a canonical way of doing calculus on Euclidean space. And, as a subset of eight-dimensional Euclidean space, there is a well-defined way of doing calculus on the seven-sphere. One defines rates of change of functions and derivatives of other objects, just by considering them in the larger, eight-dimensional space.

Abstractly, however, in precisely the same spirit in which Riemann had noted that geometry stemmed from an additional structure on a space that defined distance, all that one needs to do calculus is a consistent definition of first-order change: what it means to be linear. Two observers have to agree on what is straight. Any such agreement is called a *differentiable structure*. Milnor discovered that there were distinctly different differentiable structures on the seven-sphere.[206] There could be two seven-spheres that were homeomorphic, that is, which could be put in a one-to-one continuous correspondence with each other, but that could not be related by a one-to-one map whose rate of change was everywhere defined and not zero. Instead of only one way of doing calculus on the seven-sphere, there were twenty-eight of them. All different, all inequivalent.

Milnor's arguments were breathtaking. He brought together topology and analysis in a wholly unexpected way, and in doing so initiated the field of differential topology. In addition, Milnor is one of the most elegant mathematical writers of all time. His mathematical prose is simple, spare, and exceedingly beautiful. His prose style is to mathematics what Hemingway's is to English or Simenon's to French. His paper detailing the different differentiable structures on the seven-sphere is a mere six pages.[207] It is full of luminous insights. He used a deceptively simple technique to determine whether a manifold was a sphere[208] and studied the differentiable structure by studying eight-dimensional manifolds whose boundaries were the spheres in question.

Milnor's discoveries unleashed a flood of work, each result more spectacular and more surprising than the last. A few years later, Stephen Smale, of Berkeley, used arguments that had been started by Poincaré, but that had been perfected by American topologist Marston Morse, Milnor, and the Russian topologist Lev Pontryagin, to prove the analogue of the Poincaré conjecture for all spheres of dimension five or greater. More precisely, he proved that in any dimension *n* greater than four, a simply connected *n*-dimensional manifold that does not have a boundary and that does not go on forever, and that has the same homology as the *n*-dimensional sphere, is an *n*-dimensional sphere. (Remember that *simply connected* means that every loop can be shrunk to a point: in Poincaré's language, the fundamental group is the identity.) Poincaré had asked whether every simply connected three-dimensional manifold that does not have a boundary and that does not go on forever is a three-dimensional sphere.

The reason the statement of the original Poincaré conjecture seems simpler than the analogue that Smale proved is that Poincaré showed that simply connected in dimension three implies that homology of the manifold is the same as that of a sphere. This is not true when the dimension is greater than three, and needs to be assumed.

Smale also proved a major result that related the properties of two manifolds that bounded a manifold of one dimension higher. This pushed Thom's work even further. Christopher Zeeman in Britain, and Andrew Wallace and John Stallings in the United States produced very different proofs. Mathematicians found manifolds that had no differentiable structure,[209] and lots of other manifolds that admitted many different differentiable structures.

Higher dimensions had arrived. One might think that higher-dimensional versions of three-dimension results and conjectures would be harder to establish. Three dimensions are definitely harder to handle than two. Adding more dimensions made things more difficult to visualize. The sheer variety of different manifolds and behavior multiplies dramatically as dimension increases. But the saving grace is that there is more room: What one loses in geometric intuition is more than adequately compensated by the extra room afforded to approximate badly behaved functions and mathematical objects arbitrarily closely by simpler objects that are nicely behaved. Kinks in a manifold can be smoothed out, and critical points in functions can be slid around one another and often canceled.[210]

Smale's methods dealing with spheres of dimension five and higher broke down completely in dimension four, not to mention dimension three. The Poincaré conjecture for spheres of dimension four was proved twenty years later in 1982 by Michael Freedman, of the University of California at San Diego, now at Microsoft, using totally different techniques. He was able to classify all simply connected compact four-dimensional manifolds. Freedman had worked for eight years on his result. Rob Kirby, of Berkeley, commented to *Science*, "I think it is one of the loveliest pieces of math I have ever seen. It has an element of originality. If Freedman hadn't done it, I don't think anyone would have done it for a long time." When Freedman's techniques were combined with equally sensational work of Oxford's Simon Donaldson, who was investigating certain equations motivated by physics that existed on all four-dimensional manifolds, even more amazing results flowed. It turned out that there were infinitely many inequivalent differentiable structures on four-space! Put differently, there are infinitely many, incompatible ways of doing calculus in four-space. This contrasts with every other dimension: For any dimension except four, there is only one differentiable structure on the space underlying the Euclidean space of that dimension (that is, on the space of n-tuples of real numbers, n any positive integer except four).

In the midsixties, Milnor's "exotic spheres," as spheres with nonstandard differentiable structures were called, were discovered to occur near singular points of sets defined by extremely simple equations. Singular points are points at which all derivatives (that is, rates of change) were equal to zero, and had long been of interest to mathematicians. Stunned, mathematicians from almost every country in Europe, and from Vietnam, India, Australia, Canada, Brazil, the Soviet Union, and the United States began using techniques from differential and algebraic topology to study solutions of various types of equations. Milnor wrote up some of the results applying to polynomials, and added many more, in a short, exceedingly elegant book that became an instant classic.[211] Thom applied topology to study change in biological processes.[212]

Chemists and physicists began to use topology to study crystal defects and other phenomena. In the 1980s, topology appeared increasingly in theories aimed at linking quantum mechanics and general relativity.

THE THREE-DIMENSIONAL POINCARÉ CONJECTURE

Maddeningly, despite the deluge of mathematical discovery, the original Poincaré conjecture remained unsolved. Manifolds of dimension greater than three had once seemed esoteric. By 1982, one knew that any manifold of dimension greater than three that shared the most obvious properties that a sphere possessed was, in fact, a sphere. But, forget about this higher-dimensional stuff. What of dimension three? The universe is a three-dimensional manifold. We live in it. If every loop in it can be shrunk to a point, is it a sphere?

It is hard to imagine a simpler question about the universe. Unlike the new discoveries about higher-dimensional manifolds, it does not require complicated mathematical machinery to understand. Even worse from a psychological point of view, there were several approaches to the Poincaré conjecture that have an obsessive draw and that promised to reward effort.[213] By 1960, a number of mathematicians had spent twenty or more years working on the Poincaré conjecture. They had proved many things, but no one was able to prove it true, or false.

This was not to say that there was not progress. Ralph Fox, Milnor's advisor, and one of the leading knot theorists of the time had invited a then-unknown mathematician Christos Papakyriakopoulos (1914–76) to Princeton. Papa, as he was known, was born in Athens as the First World War broke out, and received his Ph.D. from the University of Athens in 1943.[214] He joined a guerrilla group seeking to oust the Nazi occupiers in 1944, taught primary school while in hiding in the countryside, and was forced to leave in 1946 as civil war erupted. All the while, Papa continued to work on low-dimensional topology. He sent Fox a purported proof of Dehn's lemma, the result that Dehn had thought he had proved in 1910, but that was found to contain an error nineteen years later in 1929. Neither Dehn, who had narrowly escaped Germany on the trans-Siberian railroad,[215] nor anyone else had been able to repair the proof. Fox found an error in Papa's proof, but was very favorably impressed. At Fox's urging, Papa left for Princeton in 1948 never to return to

Greece, except briefly for the death of his father in 1952. The Greek Security Police pursued him to the United States, trying to convince American immigration authorities to expel him. Princeton supported him, providing a small stipend and an office.

Princeton's generosity was amply repaid. Papa proved a critical result called the *loop theorem* in 1957, followed by a strikingly ingenious (and correct) proof of Dehn's lemma.[216] Papa had used a new construction, now known as the *tower construction*, that neatly circumvented previous difficulties. The graduate students at Princeton at this time had amused themselves by writing limericks about the mathematicians in the department. It was Milnor who supplied this one about Papakyriakopoulos:

> The perfidious lemma of Dehn
> Was every topologist's bane
> 'Til Christos Papa
> Kyriako
> Poulos proved it without any strain.

The last line refers to Papa's tower construction.

In the late 1950s, it seemed as if every topologist took aim at the Poincaré conjecture, determined to conquer it. R H Bing, a product of the strong American midwestern school of topology established by E. H. Moore's (and Veblen's) student Robert Lee Moore, spent the academic year beginning in 1957 at Institute for Advanced Study. He had studied analogues of Dehn surgery in which one sewed solid tori into cubes with knotted holes, thereby creating three-manifolds. It was believed that some of the resulting manifolds might provide counterexamples to the Poincaré conjecture. Bing did not entirely rule this possibility out, but showed that certain classes of knots could not give counterexamples. He almost withdrew the paper, however, as rumors swept the Institute that the conjecture had fallen. Two different purported proofs were circulating. Both had gaps that could not be fixed. One, instructed by a Japanese mathematician, made it into print in 1958 but it did not pass muster.[217] Poincaré conjecture 5, Mathematicians 0.

Papa turned all his efforts to proving that the conjecture was true, publishing some partial results in 1963.[218] Until his death in 1976, he lived a spartan lifestyle, arriving in his office early and leaving after five. He broke only for lunch, tea—when he scanned the *New York Times*, and the odd seminar.

But in the end, the conjecture defeated him. Poincaré conjecture 6, Mathematicians 0.

Papa's life was fictionalized in Apostolos Doxiadis's novel *Uncle Petros and Goldbach's Conjecture*.[219] The novel tells the story of a brilliant individual, Petros Papachristos, obsessed by a famous mathematical problem. (The problem is Goldbach's conjecture, the as-yet-unproved assertion that every even number is the sum of two prime numbers, which is much easier to explain than the Poincaré conjecture.) After a very promising start to his career, Petros pushes away all human contact as he works single-mindedly on the conjecture. In the end, he leaves his academic job and returns to Greece, where his brothers never forgive him for squandering his intellect. He dies, considering himself a failure because he did not prove the result.

The Poincaré conjecture, indeed, was a heartbreaker. By early 1960, about all that was clear was that no one had any idea whether or not it was true. At a conference in Georgia in 1961, Fox wrote a paper suggesting another way to look for counterexamples. Bing wrote a careful survey article in 1964 suggesting still other approaches.[220] John Stallings wrote a paper on how *not* to prove it.[221]

The higher-dimensional advances had, if anything, raised the stakes on the classical Poincaré conjecture. However, they also misled. Although the truth of the conjecture was very much in doubt, the overwhelming tide of opinion in 1980 held that the Poincaré conjecture was a purely topological question. Almost no one dreamed that geometry would have anything to do with it.

THURSTON

The 1970s, however, saw a rebirth of geometry largely at the hands of one individual, Bill Thurston. Thurston received his bachelor's degree from slightly countercultural New College in Sarasota in 1967, and his Ph.D. from Berkeley in 1972 under the guidance of Morris Hirsch and Stephen Smale. After a year at the Institute for Advanced Study and one year as an assistant professor at MIT, he was appointed a full professor at Princeton in 1974.

Differential geometry had blossomed in the twentieth century, partly as a result of its connections with general relativity. But geometry in the sense of Klein and Poincaré and Hilbert had not fared so well. Thurston changed all that. His was the most fertile and original geometric imagination since Riemann. Thurston thought about what it would be like to live in a three-dimensional manifold. What would we see if we lived in a three-torus populated with a number of objects? How would its size relative to ours matter? What about the speed of light relative to the size of the manifold? What would we see as someone walked away from us?

No one imagined that anything like the mind-blowing synthesis toward which Klein and Poincaré had raced in two dimensions might be possible in dimension three. Things were just too messy to even hope for an analogue in three-dimensions of the already miraculous-seeming fact that every surface has a unique natural geometry. There were far too many three-dimensional manifolds, and the only pattern seemed to be no pattern. It was hard to know where to start. There were, of course, the direct analogues of the three types of geometry in dimension two and the simply connected model spaces on which they existed. The usual metric on ordinary three-space gave it the geometry of Euclidean three-space which was flat. The three-sphere had a spherical geometry and had been described by Riemann. The interior of the unit ball in three-space had a natural hyperbolic geometry that Poincaré had described in the passage quoted in chapter 10. Dehn and some of the German topologists had realized that some compact three-manifolds carried a hyperbolic geometry, and other manifolds were known that carried spherical and flat geometries. But geometries on manifolds seemed rare and somewhat of a curiosity.

Thurston asked himself what we mean by an especially nice geometry. In dimension two, various different definitions coincided. Constant curvature is the same as having the same rule for measuring lengths and angles at all points and in all directions. In dimension three, there were several possible definitions, and they did not all agree. Undeterred, Thurston came up with a now widely accepted provisional definition and showed that there were eight, and only eight, different geometries in dimension three, as opposed to the three in dimension two. In addition to the spherical, flat, and hyperbolic geometries, some hybrid types existed on very particular spaces.[222]

As an undergraduate, Thurston had spent hours thinking of examples of manifolds, and geometries on them. In his Ph.D. thesis, he had studied ways of decomposing three-manifolds into stacked sheets that wrapped around one another in complicated ways.

To even talk of all three-dimensional manifolds having a natural geometry seemed hopeless. It was easy to construct counterexamples that showed that the sort of naïve things that one might hope to be true were false. However,

Thurston conjectured that any three-dimensional manifold could be carved into pieces, by cutting along two-dimensional spheres and tori in an essentially unique and natural way, each resultant piece of which had one of the eight geometries. He was able to show that his conjecture held for a very large class of three-manifolds. The *geometrization conjecture*, as he called it, implies the Poincaré conjecture.[223]

The geometrization conjecture laid out a sweeping vision of three-manifolds, but almost seemed too grand and far out of reach. Nonetheless, Thurston was able to show that most (in a suitably defined sense of the word) three-manifolds carried a hyperbolic structure. This was a total surprise. The same statement is true for two-manifolds: except for the sphere and the torus, all (orientable) two-manifolds have a hyperbolic geometry. Klein and Poincaré had known this, but the result is not usually phrased this way. Certainly, no one had suspected that anything like this might be true for three-manifolds.

The applications were immediate and delightful. For example, one consequence was that the region outside most knots (known as the *knot complement*) in the three-sphere possesses a metric that makes it a hyperbolic manifold. Although the knot was infinitely far away from the point of view of an observer in the manifold, the volume of the region turns out to be finite and gives a new number associated with the knot. Mysteriously, these volumes seem to be connected in some as-yet-unexplained way with number theory.

Before Thurston's work, the only reason to think the Poincaré conjecture might be true, was that no one could think of a counterexample. Worse, when one systematically tried to construct counterexamples, one would often be left feeling that there was no good reason why they might not exist. It was utterly maddening. After Thurston, there was a reason the Poincaré conjecture might be true. Perhaps all three-manifolds were built up of pieces that had a geometric structure.

Like Thom in 1958, Milnor in 1962, and Smale in 1966, Thurston was awarded the Fields Medal. This is the most coveted prize a mathematician can receive. It was set up by the Canadian mathematician, John Charles Fields (1863–1932). Fields had worked selflessly on behalf of the international mathematical community, overcoming a threatened boycott by French mathematicians if German mathematicians were invited to the 1924 congress. He had left a substantial portion of his estate to endow the prize that was established after his death—and over the strong objections of Veblen, who thought that research should be its own reward. The first prizes were awarded in 1936. Fields's will specified that the prize should be awarded to encourage young mathematicians. By tradition, this is interpreted to mean that the prize can only go to individuals less than forty years old at the beginning of the year in which the quadrennial congress takes place. The Second World War had interrupted mathematical congresses, but the awards have since been made every four years at international congresses from 1950. Cold war troubles resulted in postponing to 1983 the Warsaw congress at which Thurston received the medal—it had been originally scheduled for 1982. Because of the age limitation, this was the last congress at which Thurston could have received the award.

Thurston's prize occasioned some raised eyebrows at the time. He was the quintessential oral mathematician. It is unfair to say, as some have, that he did not publish enough. His work on foliations, for which he received the Fields Medal, was certainly thorough and carefully documented. However, he published much less of his purely geometric work. There was a single paper in the *Bulletin of the American Mathematical Society*, and a set of Princeton lecture notes that passed from photocopy machine to photocopy machine, and that are now available on the Web. The first few chapters of the notes were carefully rewritten by groups of mathematicians and appeared as an influential book edited by Silvio Levy.[224] Although Thurston wrote little, his students and his co-workers have published massive amounts.

Thurston has been far more reflective than most mathematicians about the unusual sociology of mathematics. He writes about his regrets in having inadvertently stilled research for over a decade into foliations of manifolds. Other less-established mathematicians, recognizing his ability, assumed that he would clear up the major problems and abandoned the field. They sought other areas in which to prove themselves. Instead of advancing the study of foliations, Thurston ruefully worries that he inadvertently retarded it.

In geometry, on the other hand, his influence has been prodigious. It is not uncommon to talk to geometers and topologists who have had their whole way of thinking about a set of problems altered by a conversation with Thurston. His ideas have so completely revolutionized our thinking about three-manifolds that even those who have never met him routinely use concepts and examples that he pioneered. Mathematicians have reembraced geometric ideas, and the field of three-manifold topology has seen an unprecedented influx of young researchers bringing geometric methods to bear on topological and algebraic problems.

Thurston has given more to the service of the mathematics, to teaching, and to thinking about how mathematics is taught and learned, than any mathematician at his level and his comparatively young age. His influential paper on mathematical knowledge makes more penetrating observations and raises more interesting questions than does any paper of comparable length on mathematics education.[225] Thurston's views have, in part, triggered a very lively debate in the mathematics community about proof and intuition.[226] These debates may be an end-of-the-century phenomenon and recall the exchanges a century earlier among Poincaré and Hilbert and others, regarding the nature of proof and of intuition.

HAMILTON AND THE RICCI FLOW

Thurston's work touched off an enormous resurgence in geometrical work. Geometric structures à la Thurston seemed to pop up everywhere. Invariants of hyperbolic manifolds began to play an important role in topology and algebraic geometry. No one, however, knew how to make further progress on Thurston's geometrization conjecture. A number of methods had been proposed, but the obstacles were formidable.

Some promising ideas came from analysis. In the early 1980s, a number of people began to investigate what happens when one takes a manifold with a Riemannian metric and tries to improve it by some sort of procedure that smooths extremes of curvature. For instance, if the curvature is high at a particular point in a particular direction, one could try to decrease it in that direction, and likewise if it seems low, one might try to increase it. With luck, maybe one could deform things so the curvature in all directions at that point and at nearby points becomes the same. With even more luck, perhaps whole regions of the manifold under consideration would develop one of Thurston's geometries and one might be able to see one's way through to establishing the geometrization conjecture for that manifold. The difficulty is to find an analytically tractable way to frame this idea.

In the early 1980s, Richard Hamilton proposed thinking of a manifold with a Riemannian metric as if it were made of metal, say, with varying temperature. What if one let the curvature flow from more curved areas to less curved areas, much as heat flows from warmer to cooler areas? This amounts to changing metrics on a space so that distances decrease fastest in directions along which the positive curvature is greatest.

FIGURE 50. *Richard Hamilton*

To quantify the rule that heat flows from warmer to cooler areas, one specifies that the temperature moves toward the average of the temperatures on a little sphere about the point. The operator that averages quantities on little spheres about a point is called the *Laplacian*. The *heat equation* specifies that the rate of change of the temperature with respect to time be proportional to the negative of the Laplacian. (Incidentally, precisely the same mechanism is at the root of the Black-Scholes equation, which governs the pricing of options in financial markets. The equation is nothing but the heat equation dressed up in financial garb.)

One wants an analogue of the heat equation for curvature. To write out a meaningful equation, both sides of the equation must be the same type of mathematical object. In the case of the heat equation, temperature and its rate of change are both numbers. But the rate of change (the time-derivative) of a Riemannian metric turns out to be a mathematical object that assigns a number to every planar direction through a point. Moreover, it must be independent of the choice of coordinates since the Riemann metric is. Recall from

Chapter 7 that the Riemann curvature is such an object. The number it assigns to each planar direction reflects how much angle sums of tiny geodesic triangles tangent to that planar direction tend to deviate from 180 degrees.

What other mathematical objects assign a value to each planar direction through a point and are independent of the choice of coordinates? There are essentially only two. One is the *Ricci tensor,* which is obtained from the Riemann tensor by averaging different combinations of curvatures in different directions. The other is the "scalar curvature" (the average of the curvatures in all directions) times the Riemann curvature tensor. Thus, to get an analogue of the heat equation, essentially the only choice is to specify that the time derivative of the Riemannian metric at each point be proportional to the negative of Ricci tensor. This is called the *Ricci flow.*[227] Hamilton proposed to study how manifolds evolved in accord with it as a way to get at the geometrization conjecture.[228] It was precisely these equations that Perelman wrote on the board at the beginning of his first lecture at MIT.

Hamilton's equations for the Ricci flow are a type of differential equation, known as a *partial differential equation.* A differential equation is one in which one specifies rates of change of some unknown mathematical object and seeks the object as a solution. Partial differential equations are a type of differential equations in which one specifies rates of change at different points in different directions. The solutions of partial differential equations are objects that have the desired rates of change at all points in all directions.

Most of the equations of mathematical physics are partial differential equations. Maxwell's equations unite electricity and magnetism by setting out partial differential equations that describe how the electric and magnetic fields change and interact from point to point as a function of the points and the directions of the field. The Einstein equations that link matter, the curvature of space, and gravity are partial differential equations. So are the equations that govern fluid flow and heat conduction, and the Schrödinger equation from quantum mechanics. Partial differential equations are of enormous practical importance and have been studied intensively for well over a century. Their study received renewed emphasis on account of the Second World War—the development of supersonic jet flight required an understanding of how solutions to the equations describing fluid flow around airplane wings depended on the shape of the wing and the speed and direction of the flow around it. Better hurricane prediction also requires better methods for solving these same equations.

How does one solve, or try to solve, a partial differential equation? The usual first step is to think through what the set of possible solutions might look like and to determine what sort of structure the space of possible solutions might carry. That the set of potential solutions, in our case the set of all metrics on a space, turns out to be an infinite-dimensional space is no surprise: the set of functions on the real line is already an infinite-dimensional space, each function being a point in the space. Next, one tries to interpret the partial differential equations as providing a flow on the infinite-dimensional space and tries to follow it. However, to do analysis on such spaces, one must exercise extreme caution. There are many potential pitfalls. There is a lot of room in an infinite-dimensional space, and it is easy for the path prescribed by the equation to lead right out of the space.

This happens when we start deforming a metric and wind up with something that is no longer a metric—the distances assigned by it could become zero or negative, they could go to infinity, or they could cease to be continuous. When such anomalies happen, we say that the solutions to the equations *develop singularities.* We need to be able to avoid singularities. Even assuming that the flow winds up somewhere nice, we need to be able to follow it. But, unless we can solve the equation exactly, which almost never is the case, we can never exactly follow a flow. We can only follow it approximately. Thus, we need some sort of bounds that tell us that we are not too much in error and that where we are going is generally in the right direction. Such bounds also allow us to make midcourse corrections. We need another set of bounds that guarantee we will stay away from singularities. Analysts call such things *estimates.* Estimates tell us when we are close enough to where something is going, to be able to continue following it. Analysts love them. Almost everyone else dreads them. Handling estimates requires great skill. And handling them near singularities requires great imagination as well as great skill.

Ever since Riemann, who made basic contributions to the theory of compressible flows, geometers had used partial differential equations to study deformations of various geometric structures. In an odd twist of fate, it seems as if the most challenging partial differential equations—the ones that are just past the limit of what one could solve, but not so far past that they are hopeless—are those that come from geometry. The 1970s had seen a number of dazzling successes showing that if one began with a manifold whose curvature satisfied certain constraints, one could alter the curvature continuously to make it nicer and to wind up with metrics that were especially symmetric.

Shing-Tung Yau had received a Fields Medal with Thurston in 1983 for showing, among other things, that one could find flat metrics on certain spaces by deforming an initial metric using a partial differential equation. In 1981, in a tour de force, Hamilton showed that if one started with a metric in which the curvature was never zero or negative, the Ricci flow wound up giving a metric of constant positive curvature.[229]

Hamilton's result was sensational. At a key point, he had to use the infamous *Nash-Moser inverse function theorem* that Nash used to prove that any Riemannian manifold could be imbedded in a Euclidean space of suitably higher dimension. Hamilton's argument was simplified by Dennis DeTurck, of the University of Pennsylvania.[230] In 1986, Hamilton and Michael Gage, now of the University of Rochester, were able to show that an analogous argument applied to closed curves in the plane actually resulted in the curve becoming a circle.[231] Given a curve in the plane and a point on the curve, we can define the curvature of the curve at that point to be the reciprocal of the radius of the kissing circle at that point. (The kissing circle at a point of a curve is the circle that is tangent to, and has the highest order of contact with, the curve at that point.) If at every point the curve evolved in the direction perpendicular to the curve at a speed proportional to the curvature, then Gage and Hamilton were able to show that the curve shrank and became a circle. Phrased differently, the curvature spread out, becoming constant.

This is plausible but far from obvious. And the more one thinks about it, the less obvious it seems. Suppose, for example, that we start with a curve like the one in figure 49. The regions of greatest curvature will move fastest, but it is not obvious and, in fact, it seems distinctly possible that parts of the curve will collapse on themselves. Gage and Hamilton's result guarantees that they do not.

By the early 1990s, Hamilton and his co-workers had shown that if one started with any compact two-dimensional surface and let the curvature evolve according to the Ricci flow, then one wound up with a constant curvature surface.[232] The curvature spread out until it became constant. This gave a conceptually simple proof that any two-dimensional manifold carried a unique geometry, the result over which Klein and Poincaré had labored so hard.

Alas, Hamilton was able to show that in general in the case of three-dimensional manifolds, the Ricci flow gave rise to singularities. If there were points on the manifold where the curvature was zero, the Ricci flow developed

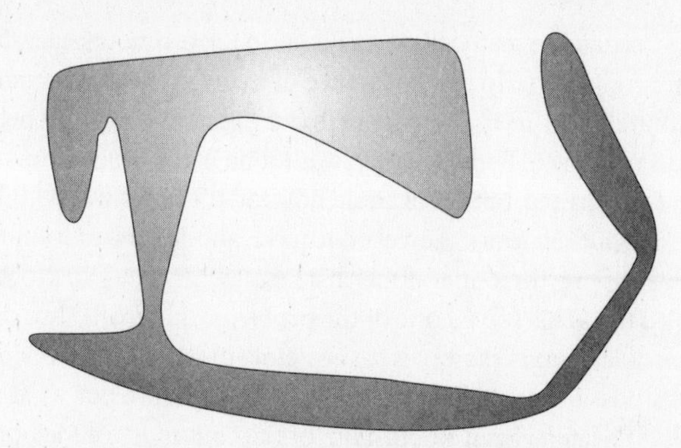

FIGURE 51. *If the curve moves at each point along the perpendicular to the curve, in the direction of the center of, and at a speed equal to the reciprocal of the radius of the circle that best fits the curve (this is the curvature), then the curve shrinks to a point, becoming more and more circular as it does so. This is the Gage-Hamilton theorem.*

horrific singularities. There seemed to be no way to avoid them. Worse, there were lots of possibilities for singularities, and while one could get estimates near some, there seemed to be no way to handle them in general. Much work continued to be done with Ricci flows techniques, particularly in higher dimensions, and it was a powerful technique for studying Riemannian geometry. But there seemed no way to use it to get at the geometrization conjecture, and hence the Poincaré conjecture, using it.

OTHER ATTEMPTS

This is not to say, of course, that other attacks on the Poincaré conjecture ceased. It seemed conceivable that a proof could come from pure algebra. Dehn and Papakyriapoulos had provided earlier reductions to algebra, but, in the early 1970s, Joan Birman, of Barnard College, reformulated the conjecture as a purely algebraic statement that seemed amenable to attack.[233] Alas, nothing came of it.

Of course, low-dimensional topologists continue to view the problem as theirs alone. Algebra, they could live with. Maybe. But they found this stuff about geometries, and especially partial differential equations, particularly distasteful. In 1986, Colin Rourke of the University of Warwick and his graduate student the Portuguese mathematician Eduardo Rego announced a solution of the Poincaré conjecture. To the annoyance of many in the mathematical community, Rourke went to the newspapers before completing the peer review process. At a meeting in Berkeley, a gap was found in the proof. Rourke insisted that it could be repaired, but alas it could not, and the proof was withdrawn.

As a result of their work, however, Rourke and Rego had found an algorithm for identifying three-manifolds that result when one glues together two solid *n*-holed tori. This solved one of the problems that made it so difficult to identify possible counterexamples to the Poincaré conjecture. Previously, one could find a possible candidate, but not be able to figure out what manifold resulted. A little later, Hyam Rubinstein, of the University of Melbourne, discovered a different algorithm. In 1994, Abigail Thompson, of the University of California at Davis took a newly developed concept called *thin position* and used it to revolutionize the problem of recognizing the three-sphere by proving, reinterpreting, and refashioning Rubinstein's algorithm. Rourke combined the Rego-Rourke algorithm and the Rubinstein-Thompson algorithm to write a computer program that could be used to search for counterexamples to the Poincaré conjecture. Although computers had gotten much faster, they still took a painfully long time to check a prospective counterexample. Also, if there were no counterexamples to the Poincaré conjecture, the process would never end. Nonetheless, a couple of graduate students implemented a computer program and began to run it.

There was a brief flutter in 1995, when Valentin Poénaru sketched an argument to prove the conjecture by establishing a four-dimensional result that would imply the Poincaré conjecture. Sadly, he had to withdraw one of the results on which he had relied.

And finally, one couldn't rule out differential geometric arguments. Hamilton's approach via the Ricci flow still had promise, although the technical difficulties seemed overwhelming. Another approach, of Michael Anderson's, looked at using a rule, again described by a partial differential equation, to let the *total scalar curvature* (a number at each point obtained by averaging the curvature in every direction at that point) flow from points of greater scalar curvature to lesser curvature, and vice versa. As in the case of the Ricci flow, one encountered singularities that seemed too complicated to analyze.

As the century drew to a close, the Poincaré conjecture appeared as far as ever from solution. The score had turned into a rout: Poincaré conjecture 50, Mathematicians 0.

Although the lack of progress was discouraging, Poincaré's topological ideas had invaded all areas of mathematics. Connections between topology and physics, and between topology and computer science, had begun to appear. The discoveries of new invariants of knots led to previously unsuspected invariants of three- and four-dimensional manifolds. The beginnings of an entirely different kind of geometry, called *symplectic geometry*, visible in Poincaré's last paper, matured into a fully fledged field. Poincaré's discovery of chaos was fully integrated into mathematical consciousnessness; and one of the leading problems was whether the equations governing fluid flow, and hurricane formation in particular, had genuinely chaotic regimes. Looking back over a century of unprecedented achievement, one could discern Poincaré's influence everywhere. Never had mathematics seemed healthier.

But there were also clouds on the horizon. The number of individuals studying advanced mathematics began to decline. The number of American mathematics bachelor degrees awarded in 1990 was less than half of those grants in 1975, and it fell further between 1990 and 2000. Similar dramatic declines from 1975 to 2000 showed up in the number of Americans and other nationals receiving Ph.D.s. The decline in the mathematical infrastructure in the Soviet Union was even more marked. With the collapse of the Soviet Union, the world's largest mathematical community dispersed. Although mathematics had flourished from the 1960s to the '80s, progress stuttered when some of the century's major results were not terribly well documented. Never had the future of mathematics seemed more imperiled.

14

A Solution in the New Millennium

It will be some years before historians achieve the distance required to begin to meaningfully assess the twentieth century. No century had ever seen such wholesale slaughter . . . or experienced such an explosion in knowledge. The century's turbulent history echoed in the timing of the quadrennial international congresses that brought together mathematicians from all areas of mathematics and all parts of the world: On account of the Second World War, there was no meeting after 1936 until 1950. There would be no International Congress to mark the new millennium in 2000 because a meeting had been held in 1998 at Berlin and the next was scheduled for Beijing in 2002.

Nonetheless, on May 24, 2000, the imposing buildings that housed the Academy of Science in Paris thronged with mathematicians. American philanthropists Landon and Lavinia Clay had endowed a new institute dedicated to the furtherance and dissemination of mathematical knowledge. The institute's high-level advisory group had decided that the best way to proceed with the Clay Institute's mission was to identify seven long-standing mathematical problems and to offer a prize of one million dollars for each solution.[234] Fearing that young mathematicians would address simpler problems in attempting to establish a track record of publications, the committee hoped the prize money would encourage the boldness to tackle complex problems. As one of the institute's advisors, Fields medalist Alain Connes, explained, the group believed that while the struggle to the top might be brutal, the view would be so breathtaking and revolutionary that all good things would follow.[235]

The seven problems were presented to mathematicians and the public at a special millennium meeting. Anticipation ran high. The committee members, themselves first-rate mathematicians, had consulted widely. Not everyone agreed with the strategy of offering monetary rewards, but there was a great deal of curiosity about what problems had been selected. In fact, the institute's Web site crashed under the pressure of an unexpectedly large number of hits, and the traffic on the mirror site at the American Mathematical Society threatened to bring down that society's servers.

More than any other scientists (and many would argue that mathematics is more art than science), mathematicians have a sense of history. The millennium meeting consciously echoed the Paris meeting in 1900 where David Hilbert's address to the international congress had provided a list of problems that had set the mathematical agenda for the new century. The 2000 meeting began with a short address by Clay Institute president Arthur Jaffe. He ended by playing a recording of Hilbert's famous speech of 1930, one of the last public lectures that Hilbert gave and the first to be broadcast on radio. Hilbert had been granted honorary citizenship by the town council of his birthplace Königsberg in honor of his retirement. Touched, Hilbert had carefully prepared and given a forceful address arguing that the entire culture of the times, insofar as it was concerned with the understanding and harnessing of nature, rested upon mathematics. He decried intellectual pessimism and the notion that there was such a thing as an unsolvable problem.

Almost seventy years to the day later, there was an almost palpable shiver in the Academy of Science lecture hall as Hilbert's strong, clear voice rang out and the speech ended with what was to become his epigram: "*Wir müssen wissen, wir werden wissen.* (We must know, we shall know.)" All knew the quote and could hear Hilbert's passion. Everyone recognized the double irony. Königsberg was totally destroyed in the war, and would fall to the Russians.[236] A few months after Hilbert's speech, Kurt Gödel had shown that it was impossible to find a system of logical axioms that would be sufficient to establish every conceivable result in number theory and not lead to some sort of contradiction. There were limits to logic.

Jaffe's address was followed by that of Fields medalist Timothy Gowers on the importance of mathematics. In a vintage display of mathematical understatement, he addressed the return on investment provided by funding mathematicians to pursue their own field according to their own sense of beauty and harmony. Then Fields medalist and former master of Trinity College Sir

Michael Atiyah and distinguished number theorist John Tate outlined the statements and background of the seven problems that had been chosen as the millennium problems.[237] All seven were well known to any serious mathematician, all were recognized as extraordinarily difficult, and all were understood to be enormously significant.

Instead of framing problems that explicitly called for the development of a whole new theory, the Clay Institute's advisory committee had chosen problems that were very specific and concrete. Although the decision to award prizes had not been uncontested, and there were a number of other problems that could have been on the list, no one could point to a problem on the list that did not belong. The committee had chosen well.

The first problem to be described was the Poincaré conjecture. Although it had been listed by every mathematician the Clay Institute had consulted,[238] the psychological effect of its inclusion was enormous. When problems go unsolved for a while, discouragement can set in. Perhaps the problem is undecidable in the sense that it requires stronger set-theoretic axioms than those typically used. Or maybe the conjecture holds, but there is no good reason why: The fluke of algebraic cancelation that would have led to a counterexample does not happen, and there is nothing more to say. Or maybe the problem wasn't really all that important anyway. Perhaps the Poincaré conjecture is just a tiny piece of the tiny, overspecialized subfield of mathematics that is classical three-dimensional topology.

The inclusion of the Poincaré conjecture sent a very clear message: The conjecture matters. It matters to all mathematicians. To all scientists, and to all of us. It is a part of our common intellectual heritage. And the tone of the conference, the broadcast of Hilbert's final words, said more. Hilbert mistakenly thought that there were no undecidable problems, but his fundamental optimism was right. We *can* settle the Poincaré conjecture. If not you or I, then someone, somewhere. And that solution will enrich all of us.

Is Grigory Perelman the One?

On November 11, 2002, Grigory Perelman posted a paper to www.arXiv.org, the online preprint server that has become the standard exchange of papers for many areas of physics, mathematics, and computer science. Perelman e-mailed a few people alerting them to the first posting. He posted two other papers four and eight months later.[239]

The November paper immediately attracted attention. First, the paper is exceptionally lucid and matter-of-fact. Addressed to those who worked on the Ricci flow, it begins with a very quick sketch of Hamilton's work: "Hamilton discovered a remarkable property of solutions . . . [which allows him] to compare the curvature of the solution at different points and different times. These results lead Hamilton to certain conjectures on the structure of the blow-up limits in dimension three . . . ; the present work confirms them."[240] What? Those very technical conjectures were well known to experts on the Ricci flow. They were exceedingly difficult, and pertained precisely to the areas where Hamilton and others had hit seemingly insurmountable roadblocks in their attempt to establish the geometrization conjecture. Establishing them would be a tremendous advance, and the geometric consequences would be astounding.

In case of misunderstanding, Perelman elaborated: "implementation of [the] Hamilton program would imply the geometrization conjecture for closed three manifolds,"[241] and went on to say that while he could not confirm Hamilton's hope that curvature does not become infinite in some regions as time goes to infinity, he could show that such regions would collapse in a controlled way, and that this is enough to draw topological conclusions.

No mathematician could fail to recognize the ring of genuineness. Then, incredibly, it became clear that Perelman was after more. Much more. He alluded to a connection between the Ricci flow and a very different flow in quantum physics that connects space at different resolutions. Here, the parameter is not time, but scale—and our space is modeled not by a manifold with a metric, but by a hierarchy of manifolds and metrics connected by the Ricci flow equation. This sort of fundamental shift in point of view was reminiscent of Riemann's probationary lecture. This mathematics belongs squarely to the new century and the new millennium, but the notion of a hierarchy of metrics would have pleased Riemann.

Perelman wrote: "Note that we have a paradox here: the regions that appear to be far from each other at a large distance scale may become close at [a] smaller distance scale; moreover if we allow Ricci flow through singularities, the regions that are in different connected components at a larger distance scale may become neighboring . . ."[242] This is the stuff of science fiction.

FIGURE 52. *Grigory Perelman*

Then, back to earth. He wrote, "Anyway, this connection between the Ricci flow and the RG [renormalization group] flow suggests that Ricci flow must be gradient-like; the present work confirms this expectation."[243] Well, almost back to earth. Gradient flows are relatively well understood, but to say that the Ricci flow can be regarded as a gradient flow represented another fundamental insight. Perelman outlined his paper, noting that the first ten sections apply in any dimension and with no assumptions on curvature. The last three pertain to the Hamilton approach to the geometrization conjecture. "Finally, in §13 we give a brief sketch of the geometrization conjecture."[244] He promised a second paper shortly with fuller details.

In effect, Perelman as much as said, "I have just proved almost everything that Richard Hamilton has conjectured about the Ricci flow. Oh, by the way, this means that I've proved the geometrization conjecture and, hence, the Poincaré conjecture. But what is really interesting is that I have proved that the Ricci flow has some properties valid in all dimensions that no one has suspected before, and these have some amazing consequences."

This was an exceedingly odd mix of understatement and boldness. Almost anyone else would have begun by saying "I've proved the Poincaré conjecture," or "I've proved the geometrization conjecture." As it was, very few people would have immediately appreciated what Perelman was driving at. To experts on the Ricci flow, however, the announcement was staggering. The technical points he mentioned were the whole ball game, but he had his eyes fixed on something far beyond the game. Perelman knew that proof of the Poincaré conjecture was staggering news, but it was almost as if he wanted to downplay it. Almost no one would recognize the import of his announcement, and those who did were those most likely to reserve judgment.

Perelman's paper contrasted with an announcement several months earlier by Martin Dunwoody, of the University of Southampton, that he had discovered a proof of the Poincaré conjecture based on the Rubinstein-Thompson algorithm. Rourke found an error, and Dunwoody withdrew the paper. Dunwoody is a highly respected mathematician. But Perelman's paper was of a completely different order of magnitude—it was far more ambitious, and the Poincaré conjecture and even the geometrization conjecture were not the main goal of the paper.

Any paper with the mix of vision, care, and authority of Perelman's would have attracted attention sooner or later. But Perelman was not exactly an unknown. Forgotten perhaps, but not unknown.

As a youth, Perelman had won the All-Union Soviet Mathematical Olympiads. The overall level of Soviet mathematical instruction was high from pupils' early childhood through graduate school, and research mathematicians were involved with the elementary and high school curricula. Students with talent were identified early, and there was an exceedingly strong tradition of mentorship. Perelman had been a student at a famous high school in Saint Petersburg specializing in mathematics and physics.[245] In 1982, he was one of three individuals to receive perfect scores on the International Mathematical Olympiad in Budapest.[246]

In the early 1990s, Perelman spent a number of postdoctoral years in the

United States, where his brilliance was noticed and is still remembered. By 1993, at age twenty-seven, he had many accomplishments: He had clarified the theory of manifolds with curvature bounded away from zero. He also had solved a major problem in Riemannian geometry, called the *soul conjecture*, that involved characterizing manifolds where the curvature was allowed to be zero.[247] If the curvature was always positive, it turned out that the manifold was homeomorphic to Euclidean space. Perelman showed that if the curvature was zero in some regions and positive elsewhere, it turned out that there was a region of the space, called the *soul*, that contained all the topology of the manifold in some sense. If the curvature was never negative and if there was even one point where some curvature was positive, Perelman showed that the soul consisted of a single point and the manifold had to be homeomorphic to Euclidean space.[248] The paper appeared in 1994, the year that he submitted his last published papers. That year, he was invited to speak at the International Congress of Mathematicians in Zurich. Afterward, he returned to Russia and seemed to drop out of sight. In 1996, he was awarded a prize by the European Mathematical Society but never showed up to claim it.

Perelman's expertise lay precisely in the area where attempts to use differential geometric methods to attack the Poincaré conjecture got stuck. His work in the early 1990s had dealt with manifolds that have regions where the curvature is zero—precisely the regions where the Ricci flow developed singularities and the analysis broke down.

And there was one final portent. Perelman belonged to the Mathematical Physics group at the Saint Petersburg Department of the Steklov Institute.[249] This is a legendary group that has made decisive and fundamental contributions to our understanding of partial differential equations. For decades until her death in 2004, the animating spirit had been that of Olga Ladyzhenskaya, the beautiful and brilliant mathematician whose father had been executed without trial by Stalinist authorities[250] and who had dedicated her entire life to mathematics. Few places would have such a concentration of individuals more likely to understand the subtleties of the behavior of solutions of nonlinear parabolic differential equations, the class of equations to which the Ricci flow belonged. Even fewer would have been led by an individual more talented, more compassionate, and more likely to understand total dedication.

Reaction to Perelman's quiet bombshell was not long in coming. The electronic mailboxes were soon overflowing. Eight days after Perelman's November 2002 post, Vitali Kapovitch, of the University of California at Santa

Barbara, e-mailed Perelman: "Hi Grisha, Sorry to bother you but a lot of people are asking me about your preprint 'The entropy formula for the Ricci. . . .' Do I understand it correctly that while you can not yet do all the steps in the Hamilton program you can do enough so that using some collapsing results you can prove geometrization? Vitali." A day later came the reply, "That's correct. Grisha."[251]

Skepticism mixed with hope. Geometers and analysts awaited Perelman's next paper, which was to supply the details filling out the proof sketched in section 13 of his first paper. The relentlessly technical paper was posted on www.arXiv.org on March 10, 2003. In it, Perelman corrected the statement of two results in the first paper, showing, however, that the corrections had no effect on the conclusions. The following month, Perelman visited the United States, giving the lectures in Cambridge and Stony Brook mentioned in chapter 1. On his return to Russia, he posted his third paper on July 17, giving a further analytic result that allowed him to use the first, and less difficult, half of his second paper to directly prove the Poincaré conjecture.[252] A month later, Tobias Colding and William Minicozzi found a still simpler, more geometric proof of this analytic result.

During the three years following Perelman's first post, his work received unprecedented scrutiny. Bruce Kleiner and John Lott, of the University of Michigan, started a Web site containing detailed commentaries on Perelman's papers.[253] Handwritten notes of his lectures were posted.

The Clay Institute, with its mission to support and disseminate mathematics, immediately swung into action. On November 2003, Richard Hamilton received a research award from the institute for his work on the Ricci flow. Lott and Kleiner, who had written up notes on Perelman's first paper, were given support by the institute to write up a very detailed tutorial expanding on almost every line of Perelman's second paper.[254] In August 2004, the institute initiated and conducted a one-week workshop in Princeton, with nearly a dozen individuals closely familiar with Perelman's work. The institute's four-week summer school in 2005 was devoted to the Ricci flow. The school featured lecture series aimed at graduate students, and proceedings are promised soon. Princeton[255] mathematician Gang Tian and Columbia mathematician John Morgan received partial support to facilitate progress on a forthcoming book on Perelman's work.[256]

Research seminars sprung up around the world to work through Perelman's results. Groups in Grenoble, Trieste, and Munich clarified many details.

In June 2005, Gerard Besson presented Perelman's work at the celebrated Bourbaki seminar in Paris.[257] Following a year-long seminar at Harvard, Huai-Dong Cao and Xi-Ping Zhu[258] wrote a long paper that explained many features of Hamilton and Perelman's work and that offered a different approach to some of Perelman's estimates.[259] Other mathematical institutes, in the United States and in Europe, and mathematicians throughout the world have played an enormous role in understanding and reworking Perelman's insights.[260]

These folks have afforded the rest of us a much fuller understanding of Perelman's work. Hamilton had classified singularities of the Ricci flow and begun a preliminary analysis of them. But singularities are things that mathematicians try to avoid. Instead, Perelman ventured deep into the regions near singularities of the Ricci flow. He found unexpected regularities when the curvature got so large that space in the manifold threatened to dissolve, and he introduced new mathematical tools to measure potential collapse. He showed that one type of singularity could not occur at all, and that others behaved in a very controlled way. In fact, the profound geometric nature of the flow became most apparent near singularities. Perelman showed that, as the flow ran, the spots where singularities occurred resulted in pieces that could be cut out of the original manifold and that had homogeneous geometries in the sense of Thurston. After the pieces were cut out, one could restart the Ricci flow and let it run until new singularities formed and with them new regions with homogeneous geometries. One could again cut out these regions and start the flow anew.

No closer interplay could be imagined between geometry and topology. It was almost as if the Thurston geometrization conjecture and the Ricci flow had been designed together. The Ricci flow was a machine that processed the manifold, stretching and shaping it, cutting off pieces with homogeneous geometries. In the end, the entire manifold had been decomposed into geometric pieces. Expositions of Perelman's work by John Morgan, Michael Anderson, and Laurent Bessières reveal the uncanny way in which the Ricci flow actually accomplishes the division of a three-manifold into pieces that carry a homogeneous geometry.[261]

The final result could not be more gratifying. Consider a three-manifold that has no boundary and does not go on forever. One can use standard methods in differential topology to give it a geometric structure. Now consider the Ricci flow and let the manifold evolve in accordance with it. If the manifold is simply connected (that is, if it is such that every loop can be shrunk to a point), then Perelman proves that the Ricci flow, after perhaps some harmless surgeries, will eventually smooth out the extremes of curvature, giving a manifold with constant positive curvature homeomorphic to the original manifold. Arguments that have been known for a long time show that a simply connected manifold with constant positive curvature is necessarily the three-dimensional sphere. Therefore, Perelman's work proves the Poincaré conjecture.

AND WHAT OF THE SHAPE OF THE UNIVERSE?

In Canton, in the very northern part of New York State, Thurston's former student and MacArthur "genius" fellowship award winner Jeff Weeks has been collating investigations aimed at determining the shape of the universe. Weeks is one of the new generation of mathematicians, who uses the Internet to work far from major research centers. He maintains a Web site, www.geometrygames.org, where a number of programs allow the viewer to pilot a spaceship in differently shaped universes.

If the universe is not too big, compared to its age, and is finite but has no boundary, then we ought to be able to see around it. There are complications because light travels at finite speed, and hence looking far out is looking back in time. But we ought to see multiple images of the same galactic superclusters. Trying to pair up different regions in the sky with one another leads to a gigantic statistical problem called *cosmic crystallography*. Closed loops that cannot be shrunk to a point would show up as spikes in the *pair separation histogram*, a mathematical tool used to search for periodicity in data. If the universe is small enough, the absence of such spikes would suggest that our universe is simply connected.

In a sufficiently small universe that is not simply connected, there is another set of data coming from what are called *last scattering surfaces* that would, under some circumstances, allow us to deduce the universe's shape. The last scattering surfaces will intersect themselves, and faint circles would appear along their intersections after averaging out background noise. The disposition of these circles would allow us to deduce the shape of space. The first years of the twenty-first century saw a search for the circles that would indicate that the universe has the topology of Poincaré's dodecahedral space.

However, the expected circles did not seem to be present. Either the universe is too big, or it is a sphere, or its fundamental group is different altogether. It is also possible that some sources of "noise" could be obscuring the circles.

At present, a number of astronomical observations suggest that the average curvature of our universe is very close to zero.[262] The preponderance of opinion among astrophysicists favors a flat universe, although one cannot rule out a universe with a slightly positive curvature. (The experimental evidence seems to exclude the possibility that the universe has negative curvature.) Because of Perelman's work, we know that if there are only a finite number of inequivalent closed loops in the universe, then it must have positive curvature. The question of the shape of the universe, however, is still very much open.

Riemann introduced manifolds as a mathematical model to explore different regions of space. In his address in 1854, he stated that there ought to be other models. Fifty years later, Poincaré completed his work on algebraic topology, leaving us with the Poincaré conjecture. Almost a century later, Perelman has left us with a gift comparable to what we received from Poincaré and from Riemann. There is no question that the surface of the Earth and the universe appear to be manifolds when viewed from a certain scale. But look more closely at the surface of the Earth and we see that there are bridges, natural and constructed, that change its topology. Zoom in even closer, and its surface stops being smooth and becomes discrete, made up of different atoms and particles. Likewise, the universe could be multiply connected with three-dimensional handles near black holes. Zoom in closer on the space itself and it appears to be a sort of quantum foam, possibly with very small higher-dimensional spheres attached to every point. Such an object with different topologies at different scales is surely better modeled by quantum mathematical objects of the sort that Perelman has glimpsed. Riemann would have approved.

THE PRIZE

The first few years of the new century have made it clear that the way in which the mathematical community operates will be very different from what mathematicians have become accustomed to over the last few decades. The organizers of the Clay Institute assumed that any serious mathematician would submit his or her results to a journal. In fact, the original bylaws insisted that solutions eligible for a millennium prize had to appear in a refereed mathematics journal of worldwide repute and have general acceptance two years later. Although Perelman has not submitted his work to any journal for more than a decade, he has certainly made his work available for examination by others. Now, the institute's bylaws accept publication in other forms.

Given that Perelman seems not to care about the prize, and had in fact started working on geometrization in 1995, well before the millennium prizes were announced, one can ask whether the millennium prizes played any role in the solution of the Poincaré conjecture. Oddly enough, the answer seems to be yes, but not in the way that the organizers imagined. A result as complex as Perelman's probably would not have won such rapid and widespread acceptance without timely support from the Clay Institute. The institute did not support Perelman—rather, it supported those who could understand his work and teach others. Refereeing is critical in mathematics. Assurance that a work is true allows others to build on it and to refashion it; however, mistaken acceptance of a result that is false can easily kill a field. Good refereeing requires understanding, and understanding means that one has to re-create the mathematics for oneself. Re-creation, happily, is easier than creation or discovery, but is still nontrivial. Perelman's work draws from many different areas and is particularly delicate. There are few individuals who could adequately assess it, and without support from the institute some of these would not have had the time to referee Perelman's work because it would pull them away from what they themselves were working on.

Nonetheless, by the end of 2005, all indications were that Perelman's proof was holding up. At a gathering convened in Trieste in June 2005 to examine progress in three-manifolds in the light of Perelman's work, participants worked through Perelman's papers and voted by acclamation that the Poincaré conjecture had been solved.[263] This was not, of course, the same as careful independent reviewing, but it was a hopeful sign. The number of false proofs previously proposed for the conjecture had made everyone cautious, and arguments using partial differential equations were recognized as especially difficult.

Then, unexpectedly, in June 2006 rumors began to swirl on the Internet suggesting that there might be some problems in Perelman's papers. Translations of Chinese newspapers implied that there were some gaps in his proof of the geometrization conjecture, but that they had been filled by mathematicians

Huai-Dong Cao and Xi-Ping Zhu.[264] Field medalist Shing-Tung Yau was quoted as saying at a large conference in Beijing on string theory that Cao and Zhu's work was absolutely essential because the gaps in Perelman's work were quite large. Confusion set in as contradictory accounts spread. On June 25, the Clay Institute posted, with no comment, links on the home page of their Web site to the Cao-Zhu paper, the Morgan-Tian book, and the Kleiner-Lott preprint, together with a link to the Web site that Kleiner and Lott maintained and to Perelman's papers. What was going on? Had Perelman's arguments collapsed? Was the Poincaré conjecture in doubt?

15

Madrid, August 2006

The twenty-fifth International Congress of Mathematicians opened on August 22, 2006, in Madrid. Nearly four thousand mathematicians from over 120 countries had streamed into Spain's capital over the preceding days.

Never had a mathematical meeting been more widely anticipated. For months, official press releases had hinted that the status of the Poincaré conjecture would be clarified during the congress.[265] Richard Hamilton had been invited to give the first plenary address, and the schedule listed a lecture by John Morgan on the Poincaré conjecture. The summer's events had only sharpened the anticipation. Throughout the summer, concern had grown about whether the conjecture had really been solved and whether Perelman's arguments were correct. Would he get a Fields Medal? The rules specified that recipients must be less than forty years of age on January 1 of the year in which the prize is awarded. Perelman had turned forty on June 13, 2006, so this would be the last time that he would be eligible. Would he show up? And, would the Fields committee award a prize to someone who rejected the prize?

The story had spilled over into the press. Articles on the conjecture and Perelman appeared in newspapers around the globe. As the conference began, *The New Yorker* published a sensational article by two respected journalists alleging that Shing-Tung Yau had deliberately attempted to cast doubt on Perelman's results in China in order to claim credit for his students.[266] The journalists had flown to Saint Petersburg, visited Perelman, and reported that the reclusive mathematician would reject the Fields medal.

The meeting was the first international congress to be held in Spain, and Madrid was an especially fitting location. Nine centuries earlier, barely forty miles away in neighboring Toledo, Gherard of Cremona's translations of Arabic works of mathematics and science, and of Arabic translations of Greek scientific works, into Latin had begun the flow of learning into Europe. His translations of Euclid, of Ptolemy, and of al-Nayrizi's commentaries on the *Elements* sparked the founding of the first universities.

Spanish mathematicians had pulled out all stops to welcome their colleagues. As they proudly noted, the meeting was the largest gathering of mathematicians in Madrid since 1581, when the then twenty-year-old decision to move the Spanish court to Madrid had resulted in the largest concentration of mathematicians in Europe. King Juan Carlos would preside over the opening ceremonies at which the Fields medals would be awarded. The popular monarch had guided the establishment of democratic government following the death of dictator General Franco in 1975. His courage in resisting an attempted right-wing coup in 1981 had been a turning point in Spain's recent history. The years since had seen tremendous economic growth and a veritable blossoming of mathematical research.

Long lines formed outside the conference building as the mathematicians stepped through metal detectors and had their bags screened. Tow trucks removed cars parked adjacent to the conference center, a precaution taken in the wake of the 2004 Al Qaida bombing of commuter trains in Madrid that killed more than two hundred and injured another fifteen hundred.

At 10:30 AM, the congress opened with a video, *Shape Through Time*, recalling the Moorish fascination with regular patterns and the geometric traditions of Spain's multiethnic past. The video was followed by the traditional short concert, this time a string trio led by Ara Malikian, Spain's gifted Lebanese-Armenian violinist, that incorporated a flamenco guitar. Finally, the president of the International Mathematical Union, the distinguished applied mathematician Sir John Ball, addressed the assembly: "While celebrating this feast of mathematics, with the many talking points that it will provide, it is worth reflecting on the ways in which our community functions. Mathematics is a profession of high standards and integrity. We freely discuss our work with others without fear of it being stolen, and research is communicated openly prior to formal publication. Editorial procedures are fair and proper, and work gains its reputation through merit and not by how it is promoted. These are the norms operated by the vast majority of mathematicians.

The exceptions are rare, and they are noticed." Strong words that most took as a commentary on the events surrounding the Poincaré conjecture reported in the press.[267]

After addresses by various dignitaries, Ball announced the four Fields Medals in alphabetical order. The second went to Grigory Perelman, "for his contributions to geometry and his revolutionary insights into the analytical and geometric structure of the Ricci flow."[268] A large picture of Perelman, looking like an Old Testament prophet, flashed before the crowd. Ball interrupted the applause: "I deeply regret that Dr. Perelman has declined to accept the Fields Medal." Some scattered clapping—signaling approval of the decision to award the medal despite Perelman's decision to decline it—followed, then silence and a barely audible collective sigh.

The awards were followed by a reception sponsored by the city of Madrid with an array of tapas. After lunch, there were short presentations on the work of the Fields medalists. John Lott summarized Perelman's accomplishments, emphasizing both their sheer novelty and Perelman's technical power.

Next, Richard Hamilton gave the first plenary address of the congress. In the extraordinarily generous talk, Hamilton surveyed his own work on the Ricci flow, the technique that he had introduced and that had been his life's work. He recounted how the idea for using the Ricci flow had come to him after hearing a lecture by James Eells four decades earlier. "The hypothesis is that there is no topology," he recalled, referring to the hypothesis that there are no loops that cannot be shrunk to a point, and hence nothing for topologists to manipulate: "So maybe we [analysts] can help them [the topologists] out." Hamilton, the analyst, could be forgiven his delight in analysis coming to the rescue of topology. One hundred years earlier, Poincaré had invented algebraic topology to rescue analysts, helpless in the face of the chaotic behavior that arose in the equations that governed planetary motions. Perelman's use of analysis to resolve the greatest problem in topology repaid, with interest, a century-old debt.

Hamilton elaborated on the key notion of entropy that Perelman had introduced to prove the critical non-collapsing theorem that allowed one to use the Ricci flow to carve a manifold into geometric pieces. Although he and his colleagues had subsequently discovered some simplifications, Hamilton emphasized that Perelman's arguments were completely correct. "I'm so grateful to Grisha for doing this," he remarked. "I had clawed by my fingernails to get [non-collapsing] in a few cases. Now I never have to worry about this again."

Hamilton's somewhat technical sketches of Perelman's proofs clarified for all in the packed hall just how breathtaking and groundbreaking those arguments were. "I think I'm surprised as anyone to see this all working," he mused. "I'm enormously grateful to Grisha Perelman for finishing it off."

Those notes of wonder and of gratitude would be echoed by others during the congress. Two days later, the chair of Columbia University's mathematics department, the accomplished algebraic geometer John Morgan, spoke on the Poincaré conjecture for the general public. Highly respected and very cautious, he had been heavily involved in checking Perelman's results. He spoke clearly and slowly so as to be understood by each person in the multilingual audience. Every seat in the huge lecture hall was occupied, and the audience hung on his every word. Lest there be any doubt, Morgan flatly announced: "Grigory Perelman has solved the Poincaré conjecture." Tension gave way to applause.

Morgan sketched the history of the conjecture, observing that it had been linked to most of the progress in geometry and topology during the twentieth century. He hailed the proof as a "stupendous achievement" not just for Perelman, but for all of mathematics. Every advance on the conjecture had given rise to substantial mathematics and resulted in Fields Medals for those involved, among them Milnor, Smale, Freedman, Donaldson, Thurston, and Yau. Morgan paraphrased Isaac Newton's self-assessment,[269] noting that Perelman had stood on the shoulders of giants, especially Richard Hamilton, whose painstaking work over twenty-five years had laid the foundations of the Ricci flow. "It's now been thoroughly checked," Morgan stated again: "He [Perelman] has proved the Poincaré conjecture." His pleasure at the resolution of the Poincaré conjecture was clear to all.

What of the Millennium Prize? Speaking to reporters in Madrid, the president of the Clay Institute, James Carlson, stated that the clock on the two-year waiting period had started ticking with the appearance of the three expositions listed on the institute's home page. Although Perelman had not published in a refereed venue, the expositions of his work had been carefully reviewed.[270] Carlson made it clear that, like the Fields Medal committee, if the institute decided to make an award to Perelman, they would do so whether or not he accepted it.

In view of the award of the Fields Medal and the reports of Lott, Hamilton, and Morgan, there can be little doubt that Perelman will be one of the Millennium Prize winners. The difficult decision of whether, and how, to split the prize will fall to the Clay's advisory board. Time will tell what their decision will be, and whether Perelman accepts this prize.

As the conference closed, more details about Perelman emerged. He had left his position at the Steklov. A Russian Jew, he lived alone with his mother; his sister and his father had emigrated to Israel. Initial reports that his declining the Fields Medal constituted a rejection of the mathematical community seem unfounded. Like Gauss, Perelman shunned the limelight and did not want to speak on behalf of mathematics. Happily, unlike Gauss, he had written up his work for the benefit of the rest of us and had made himself available to Morgan and others by e-mail to answer the occasional question calling for clarification.

The full scope of Perelman's work will become clear as the millennium progresses. Morgan speculated that Perelman's work would allow the Ricci flow to be used to investigate four-dimensional manifolds. He also thought that Perelman's techniques might be used to make progress on other types of parabolic differential equations. History has shown repeatedly that advances in handling partial differential equations lead to powerful practical applications.

Even more speculatively, Perelman's papers suggest that the Ricci flow is more than an analytic tool for obtaining geometric information about manifolds: it is a geometric object in its own right and allows one to unite manifolds with different topologies at different scales. In her lecture "Mathematical Problems in General Relativity" the day after Morgan's talk, the great French mathematical physicist Yvonne Choquet-Bruhat discussed the need for manifolds modeled on space-time which differed at different scales: the space-time analogue of what Perelman was considering in the pure space case. Perelman's work potentially provides new conceptual tools for thinking about space and time.

It has taken over a hundred years to resolve the question that Poincaré raised on the last page of the last section of the last of his great topological papers. Over those years, topology and its older cousin, geometry, have grown into coherent and powerful disciplines central to mathematics and the sciences. The mathematics on which Poincaré drew began five thousand years ago in what is now Iraq, then Babylon. The ancient mathematics was passed on and elaborated on by generation after generation, by the Greeks on the islands off what is now Turkey and their successors in Athens and Alexandria, by the Jaina and Hindu scholars in India, by the Muslim cultures east and south of the Mediterranean, by the multiethnic societies in Spain and Sicily, and then by the Judeo-Christian cultures of early modern Europe and the Enlightenment.

Poincaré's work, in turn, provided the soil for the flowering of twentieth-century mathematics. No fewer than four of the millennium problems are directly connected with work that he pioneered. We have only recently begun to fully grasp what he glimpsed at the turn of the last century.

We live in the most mathematically productive era in human history. And it is deeply satisfying to know that the beautiful work of Thurston, Hamilton, and Perelman will underwrite the work of a new era. Mathematics is the work of individuals. But its concepts and its theorems belong to no person and no ethnic, religious, or political group. They belong to all of us. Mathematical knowledge builds on the work of those who have gone before us. It is hard won, and we often do not value it as we should. Any one of us with an elementary school education can solve arithmetic and algebraic problems that would have defeated the most learned Babylonian scribes. Any one of us with a few courses of calculus and linear algebra can solve problems that Pythagoras, Archimedes, or even Newton could not have touched. A mathematics graduate student today can handle topological calculations that Riemann and Poincaré could not have begun. We are not smarter than they. Rather, we are their beneficiaries.

Mathematics reminds us how much we depend on one another, both on the insight and imagination of those who have lived before us, and on those who comprise the social and cultural institutions, schools and universities, that give children an education that allows them to fully engage the ideas of their times. It is up to all of us to ensure that the legacy of our times is a society that stewards and develops our common mathematical inheritance. For mathematics is one of the quintessentially human activities that makes us more fully human and, in so doing, leads us to transcend ourselves.

Looking up at the night sky, at the distant stars and galaxies and clusters of galaxies, it is inconceivable to me that there are not other intelligences out there, some far different than us. Hundreds of years hence, if we ever develop technologies that enable us to meet and to communicate, we will discover that they will know, or want to know, that the only compact three-dimensional manifold in which every loop can be shrunk to a point is a three-sphere. Count on it.

Notes

1. The equation $\partial_t(g_{ij}) = -2\,R_{ij}$ is the equation for the Ricci flow.

2. The only problem of comparable fame is "the Riemann hypothesis," which is also a millennium problem. The Poincaré conjecture and the Riemann hypothesis were the two problems that every mathematician consulted had listed as a potential millennium problem.

3. The institute is the Clay Institute. Its Web site www.claymath.org contains information on its mission and formation. Additional information on the establishment of the institute may be found in A. Jaffe, "The Millennium Grand Challenge in Mathematics," *Notices of the American Mathematical Society*, 53 (no. 6), 2006: 652–660.

4. For the Web site, see www.math.lsa.umich.edu/research/ricciflow/perelman.html.

5. Lucius Caecilius Firmianus Lactantius (c. AD 250–325) was a professor of rhetoric in Nicomedia who became a Christian convert, lost his job, and subsequently became a tutor of Constantine's son, and a widely known Christian apologist. For more on him, see the online edition of the 1917 *Catholic Encyclopedia* at www.newadvent.org/cathen. For current references and bibliography, consult Jackson Bryce's online bibliography of Lactantius, www.acad.carleton.edu/curricular/CLAS/lactantius/biblio.htm. The passage cited is a paraphrase in J. J. Anderson's *Popular History of the United States* (New York: Clark and Maynard, 1880) of Irving's paraphrase of Lactantius. Lanctantius did indeed believe in a flat Earth, but his was very much a minority view, and his writings were not available in Spain in 1490.

6. Washington Irving, *Life and Voyages of Columbus* (London: John Murray, 1830); new edition edited by J. H. McElroy (Boston: Twayne Publishers, 1981).

7. A superb writer, Harvard's historian Samuel Eliot Morison (1887–1976) is one of the best-known and most beloved American historians. His memorable characterization of Irving's scenes of Columbus as moonshine and malicious

nonsense come from his book *Admiral of the Sea*, vol. 1 (Boston: Little, Brown, 1942), 88–89. Oddly enough, despite Morison's rebuttal of Irving, the myth that scholars and educated individuals in the Middle Ages believed in a flat Earth is astonishingly persistent. For a wonderful account of Irving's influence and the myth's persistence, see J. B. Russell, *Inventing the Flat Earth: Columbus and Modern Historians* (Westport: Praeger Publishing, 1991).

8. The passage in the text is cited by T. Heyerdahl in *Early Man and the Ocean: A Search for the Beginnings of Navigation and Seaborne Civilizations* (Garden City, NY: Doubleday, 1979), 147. It comes from a Danish translation by C. V. Ostergaard of Caddeo, *Giornale di Bordo di Cristoforo Columbo 1492–93*, 10. Heyerdahl notes that Columbus's hypothesis makes a lot of sense in view of his beliefs that the Viking settlements represented the northern part of Asia, and that he had come to the Spice Islands east of India by sailing a shorter way across the upper northern part of a pear-shaped world instead of around a bulging large Southern Hemisphere.

9. Although there are a number of biographies of Pythagoras, the details of his life are very murky and contradictory. See, for example, P. Gorman, *Pythagoras: A Life* (London, Boston: Routledge and K. Paul, 1979).

10. For reliable accounts of Egyptian and Babylonian mathematics, consult O. Neugebauer, *The Exact Sciences in Antiquity* (Princeton: Princeton University Press, 1952) and B. L. van der Waerden, *Science Awakening I: Egyptian, Babylonian, and Greek Mathematics*, English trans. by A. Dresden with additions by the author (Leyden: Noordhooff, 1975). Neugebauer's *Mathematical Cuneiform Texts* (New Haven: American Oriental Society, 1945) and his *Vorlesungen über Geshichte der antiken mathematischen Wissenschaften con O. Neugebauer* (Berlin, New York: Springer-Verlag, 1969) contain a great deal of additional information.

11. Alexander's father, Phillip of Macedon, had completed his conquest of Greece in 338 BCE, and had engaged Aristotle to privately tutor his son.

12. Eratosthenes calculated the circumference as 250,000 stadia. If one takes 157.2 meters in a stadium (which is the value some derive from Pliny), Eratosthenes' figure is off by less than 3 percent: 24,200 miles as compared to 24,902.

13. The date of the destruction of the great library of Alexandria is uncertain and the subject of much debate. Most scholars agree that the main library was destroyed in the civil war during the Emperor Aurelian's rule in the third century CE (although others, following Plutarch, attribute it to Julius Caesar's burning of Alexandria harbor two centuries earlier). The sister library in the nearby Serapeum temple was destroyed shortly after the year 391, following the decree of the Christian emperor Theodosius.

14. A very fine English translation, *Claudius Ptolemy: The Geography*, edited and translated by E. L. Stevenson, has been republished by Dover Publications (New York: Dover, 1991).

15. The translation of the Greek text of Ptolemy into Latin was begun by the Byzantine scholar Emanuel Chrysoloras (1335–1415) and completed by his student Jacopo d'Angelo in 1406.

16. Ferdinand Magellan set out to circumnavigate the Earth in 1519 with a 265-man expedition. He was killed in battle off the Philippines in April 1521. The Basque explorer Juan Sebastián Elcano took over command and returned to Spain with seventeen survivors in September 1522.

17. A mathematician would say "boundary of some three-dimensional manifold" instead of "surface of some solid."

18. You may remember seeing, or reading about, a Möbius band. This is the object you get if you take a rectangular strip of paper and glue one short edge to the other after twisting one by 180 degrees. This is a two-dimensional manifold with boundary. It has many curious properties. You cannot consistently define right and left on it—if you start out with a stick figure on it with left and right hands defined and follow the figure around the band, it comes back with left and right hands interchanged. It also just has one side: if you start out traveling on one side and go all the way around, you come back on the other side. Thus, there is no way that it could be the surface of something.

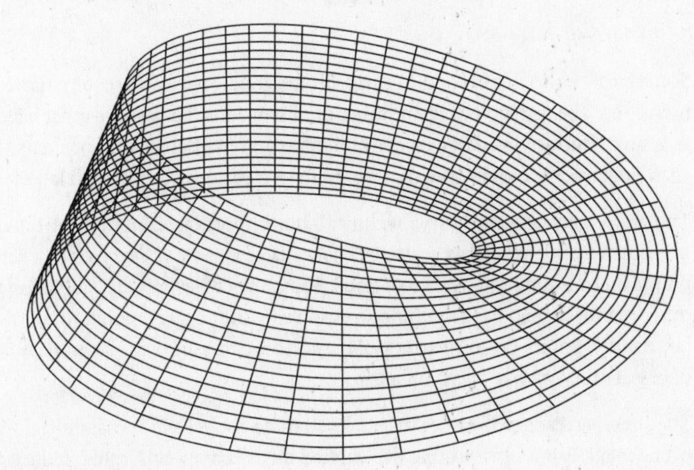

FIGURE 53. *Möbius band*

19. If a two-dimensional manifold is not orientable, it always contains a Möbius band. The boundary of a Möbius band is a single closed curve and is, therefore, topologically a circle. Attaching a disk to the band by matching the points on its bounding circle with the points on the boundary of the Möbius band gives a manifold without boundary (called the *projective plane*). Another famous example of a compact manifold without a boundary that is not orientable is the *Klein bottle*, whose inside surface is continuous with its outer surface. This is obtained by attaching two Möbius bands together by matching points on their bounding circles. The Klein bottle and projective plane are examples of two-dimensional manifolds that are not the surface of a solid.

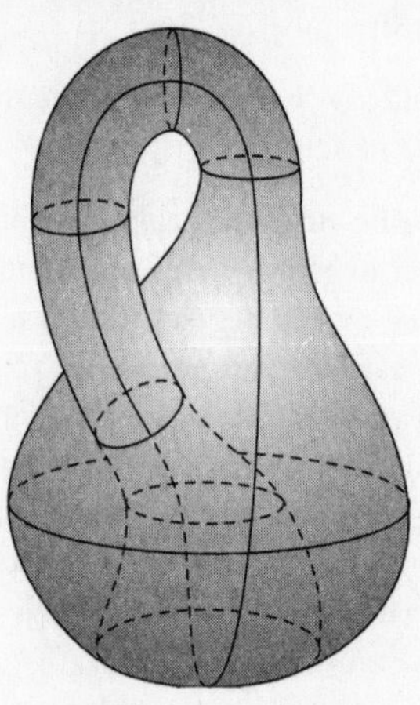

FIGURE 54. *Klein bottle*

20. Some surfaces do not even fit in three-dimensional space and require more dimensions. The question of the number of dimensions that a space must have to fit a given surface inside gives rise to interesting mathematical problems that shall not concern us here, but that we shall have occasion to mention later.

21. The word *continuous* refers to a technical, but important, condition: A one-to-one correspondence between two surfaces is *continuous* if every time a set of points on one manifold tends to a given point, the corresponding points on the other manifold tend to the point corresponding to the given point. (In particular, nearby points go to nearby points go to nearby points, but note it isn't necessary to say what "near" means.)

22. The attentive reader will notice that (at least) three issues have been lightly glossed over. First, it is not obvious that the knotted torus is homeomorphic to the standard torus. By our definition, this means that you have to be able to put the points

of each in one-to-one correspondence so that nearby points correspond to other nearby points. To see this is possible, imagine cutting each torus along a curve encircling its solid neck. In each case, you get a cylinder bounded by two circles that were the same before you cut the tori. The points on the two cylinders can be put into one-to-one continuous correspondence so that the points not on the circles correspond to one another, the points on the circles correspond to one another, and nearby points correspond to nearby points, whether on or off the circles. Second, you might ask how one knows that a sphere is not homeomorphic to a torus. The answer is again not obvious. Suppose that there were a homeomorphism (that is, a continuous one-to-one correspondence) between the two. Choose a simple closed curve (that is, one that begins and ends at the same point, and that does not cross itself) on the torus that does not divide the torus into two pieces. For example, choose a curve encircling the neck. Under the one-to-one correspondence, you would get a simple closed curve on the sphere that did not divide the sphere into two pieces. But any closed curve on the sphere necessarily divides it into two pieces. (Although this seems clear, it is surprisingly hard to prove, and is called the *Jordan curve theorem.*) Finally, we have not yet defined three-dimensional space. We do this in chapter 4 and more carefully in chapter 7.

23. The connected sum of any manifold and a sphere is the manifold you started with. This is because if you cut a disk out of a sphere, you are left with a disk, so taking the connected sum of a manifold with a sphere amounts to taking a disk out of the manifold and then sticking another disk in. Nothing changes. For the operation of connected sum, the sphere behaves like the identity element (that is, it behaves like zero for addition or one for multiplication).

24. One can also classify all nonorientable surfaces. They turn out to be connected sums of the orientable surfaces and surfaces homeomorphic to the projective plane (defined in endnote 19). For a nice statement and a modern proof, see of J. Weeks, Appendix, *The Shape of Space,* 2nd ed. (New York, Basel: Marcel Dekker, 2002).

25. V. I. Arnold, "On Teaching Mathematics," *Russian Math. Surveys* 53, no. 1 (1998): 229–36. Incidentally, this article is a wonderful rant by one of the greatest mathematicians of our times on the ineptitude of much mathematical teaching.

26. Technically, we say that a loop on a manifold is defined as a continuous map f of the interval $I = \{x: 0 \leq x \leq 1\}$ into the manifold such that $f(0) = f(1)$. We say that a loop can be *shrunk to a point* if there is a continuous map F of the rectangle $\{(x, t): 0 \leq x, t \leq 1\}$ such that $F(x, 0) = f(x)$ (the original loop) for all x, $F(0, t) = f(0)$ for all t, and $F(x, 0) = F(x, 1) = f(x)$ for all x. Then, for fixed t, $F(x, t)$ is a loop beginning and ending at the same point.

27. Argument as quoted by Simplicius, in his *Physics*. The translation is from T. L. Heath, *A History of Greek Mathematics*, 2 vols. (Oxford: Clarendon Press, 1921).

28. More technically, a one-to-one correspondence between two three-manifolds is *continuous* if every time a set of points on one manifold tends to a given point, the corresponding points on the other manifold tend to the point corresponding to the given point.

29. The analogue of the plane that Euclid envisaged is called *Euclidean three-space*, or more simply *three-space*. A more formal definition will come in chapter 7.

30. See M. Peterson, "Dante and the 3-sphere," *American Journal of Physics* 47 (1979): 1031–35, and R. Osserman, *Poetry of the Universe* (Garden City, NY: Doubleday, 1995). Apparently, this was noticed many years ago by the mathematician Andreas Speiser in his book *Klassische Stücke der Mathematik* (Zürich: Verlag Orell Füselli, 1925). The latter is referenced in the article by J. J. Callahan, "The Curvature of Space in a Finite Universe," *Scientific American* 235 (August 1976): 90–100. I owe these references to M. Peterson.

31. Map out a trip in the universe in which you stay at the same distance from the inside and outside tori, but travel in a circle perpendicular to the common axis of the concentric tori. This traverses a loop that cannot be shrunk to a point. Morevover, this loop is independent in a sense that can be made precise of the first loop. In the manifold we constructed by identifying the inside and outside boundaries of a spherical shell you do not get two loops that are independent in this way and neither of which can be shrunk to a point.

32. For a good summary of Euclid's life and work, with careful references, see I. Bulmer-Thomas, "Euclid," in C. C. Gillespie, ed., *Dictionary of Scientific Biography* (New York: Scribner, 1971). See also J. J. O'Connor and E. F. Robertson's article in the Web site hosted by the University of St. Andrews (www-history.mcs.st-andrews.ac.uk/Mathematicians/Euclid.html). This Web site features online biographies of many mathematicians, and a number of articles on special topics in mathematical history. It has gotten increasingly better and more scholarly over the years, and is now a very useful source.

33. Adapted from T. L. Heath's monumental translation *The Thirteen Books of Euclid's Elements* (New York: Dover, 1956).

34. As we shall see, centuries later, at the end of the nineteenth century, in his rewriting of Euclid, the great Göttingen mathematician Hilbert added a number of "betweenness axioms" to address precisely this gap.

35. See, for example, J. Gray, *Ideas of Space: Euclidean, Non-Euclidean, and Relativistic* (Oxford, New York: Oxford University Press, 1979).

36. L. Russo, *The Forgotten Revolution: How Science was Born in 300 BC and Why It Had to Be Reborn*, trans. Silvio Levy (New York: Springer, 2003).

37. The books *Data, On Division, Optics* and *Phaenomena* have survived. *Surface Loci, Porisms, Conics, Book of Fallacies*, and *Elements of Music* have not.

38. Both these translations were revised and others were made. Only the second, edited and almost certainly vastly revised by al-Nayrizi, survives as a manuscript in Leiden.

39. See, for example, C. H. Haskins, *The Rise of Universities* (reprinted Ithaca, NY: Cornell University Press, 1957), 1–2.

40. The inventing of the printing press is often attributed to Johannes Gutenberg in 1455. The actual story is more complex. See, for example, Adrian Johns, *The Nature of the Book: Print and Knowledge in the Making* (Chicago: University of Chicago Press, 1998).

41. See the excerpt from van der Waerden quoted in J. J. O'Connor and E. F. Robertson's article in their Web site (www-history.mcs.st-andrews.ac.uk/Mathematicians/Euclid.html).

42. Both Simson and Playfair were Scots. Robert Simson (1687–1768) became a professor at the University of Glasgow in 1710 and prepared an edition of the *Elements* containing books 1–6, 11, and 12, which went through seventy editions, the first edition in Latin, and all subsequent ones in English. John Playfair (1748–1819), a professor of mathematics, and later professor of natural philosophy, at the University of Edinburgh, published a best-selling English-language edition of the geometric books of the *Elements* systematically using algebraic notation to simplify arguments.

43. Consult J. Murdoch "Euclid: Transmission of the Elements," in *Dictionary of Scientific Biography*, ed. C. C. Gillespie (New York: Scribner, 1971). This article contains a chart filiating the different versions of the Elements in the Middle Ages. One can also profitably consult the notes in Heath's translation.

44. See "Excursus II: Popular Names for Euclidean Propositions," p. 415, in the second edition of T. L. Heath's translation.

45. These commentaries were those of Geminus of Rhodes (10–60), Heron (c. 10–75) and Pappus (c. 290–350) of Alexandria, and Porphyry Malchus (233–309).

46. To say that the fifth postulate is equivalent to Playfair's postulate means that you can prove Playfair's postulate if you accept Euclid's five postulates, and, conversely, if you accept Euclid's first four postulates and Playfair's postulate, you can prove Euclid's fifth postulate as a proposition.

47. G. Saccheri, *Euclid ab omni naevo vindicatus* (n.p.: 1733).

48. George II also founded Princeton University. Chartered in 1746, Princeton was then known as the College of New Jersey and was unique in the colonies in specifying that it was open to persons of any religious denomination whatsoever.

49. From the preface of A. Kästner, *Mathematische Anfangsgründe* (n.p.: 1758). The excerpt is quoted in W. B. Ewald, *From Kant to Hilbert: A Source Book in the Foundations of Mathematics*, vol. 1 (New York, Oxford: Oxford University Press, 1996), 154.

50. Wallis's axiom states that there exist similar figures of arbitrary size. Similar figures are polygons that have same number of sides, and the property that both the angles between corresponding sides and the ratios of lengths of corresponding sides are equal.

51. G. S. Klügel, *Conatuum praecipuorum theoriam parallelarum demonstrandi recensio* (n.p.: 1763), 16.

52. In particular, Lambert proved that the number π (the ratio of the circumference of a circle to its diameter) is not a rational number (that is, like the square root of 2, it cannot be written as a quotient of two integers). This was a great achievement.

53. I. Kant, "*Was Ist Äufklarung?*" (n.p.: 1784).

54. For an excellent biographical and scholarly assessment of Gauss's work, see W. K. Bühler, *Gauss: A Biographical Study* (Berlin, New York: Springer-Verlag, 1981). The book contains a very fine annotated bibliography.

55. The Collegium Carolinum.

56. See G. B. Halsted, "Biography, Bolyai Farkas [Wolfgang Bolyai]" *American Mathematical Monthly* 3 (1896): 1–5.

57. According to J. J. O'Connor and E. F. Robertson's online biography of János Bolyai (www-groups.dcs.st-andrews.ac.uk/history/Biographies/Bolyai.html).

58. Gauss, *Werke* (Göttingen: K. Gesellschaft der Wissenschaften zu Göttingen, 1863–1933), 8: 119.

59. Quotation attributed to Gauss by J. J. O'Connor and E. F. Robertson, www-history.mcs.st-andrews.ac.uk/Quotations/Gauss.html.

60. Quoted in V. Kagan, *N. Lobachevsky and His Contribution to Science* (Moscow: Foreign Languages Publishing House, 1957).

61. English translations of Bolyai's and Lobachevsky's essays, together with many other materials, may be found in R. Bonola, *Non-Euclidean Geometry: A Critical and Historical Study of Its Development*, trans. H. S. Carslaw (New York: Dover, 1955).

62. Gauss, *Werke* (Göttingen: K. Gesellschaft der Wissenschaften zu Göttingen, 1863–1933), 8:221. The correspondence between Gauss and Bolyai senior has been conveniently collected in Gauss, *Briefwechsel zwischen Carl Friedrich Gauss und Wolfgang Bolyai* (Leipzig: B. G. Teubner, 1899).

63. Quoted in J. J. O'Connor and E. F. Robertson's online biography of Farkas Bolyai (www-groups.dcs.st-andrews.ac.uk/history/Biographies/Bolyai_Farkas.html).

64. Three of the books are about the Riemann hypothesis, a famous problem to which we shall return later and on which the Clay Institute has also leveled a million-dollar bounty. They are: J. Derbyshire, *Prime Obsession: Bernhard Riemann and the Greatest Unsolved Problem in Mathematics* (Washington: Joseph Henry Press, 2001); K. Sabbagh, *The Riemann Hypothesis: The Greatest Unsolved Problem in Mathematics* (New York: Farrar, Strauss, and Giroux, 2002); M. du Sautoy, *The Music of the Primes: Searching to Solve the Greatest Mystery in Mathematics* (New York: HarperCollins, 2003). Derbyshire's does an especially nice job on the mathematics for a nontechnical reader. The Riemann hypothesis even figures in a recent crime novel (Perri O'Shaughnessy, *A Case of Lies* [New York: Random House, 2005]). For an extraordinarily beautiful and scholarly account of Riemann's life and work (that requires, however, substantial mathematical background), see D. Laugwitz, *Bernhard Riemann 1826–1866: Turning Points in the Conception of Mathematics* (Boston: Birkauser, 1999). (This is the English translation by A. Shenitzer from the German.)

65. The Revolution and Napoleon had further centralized instruction and administration in Paris. Even if it had occurred to them, the French have never successfully decentralized anything.

66. A. N. Whitehead, "Universities and Their Function," in *The Aims of Education and Other Essays* (New York: Macmillan, 1967), 93.

67. Its title translates as the *Journal for Pure and Applied Mathematics.*

68. The notion of number that we use today has a long and distinguished history that we cannot recount here. You can consider a *real number* to be any number—negative, zero, or positive—that can be written as a (possibly infinite) decimal: that is, as an integer plus a number between 0 and 1 with a possibly infinite decimal

expansion. The numbers 10.88901, $\pi = 3.141592\ldots$, $-1.412\ldots (= -\sqrt{2})$, 1000, 0, and -317.2 are real numbers. The complex numbers are obtained by augmenting the real numbers with the square roots of negative numbers. Every *complex number* can be written in the form $a+ib$ where $i^2=-1$ and a and b are real numbers.

69. R. Dedekind, "Bernhard Riemann's Lebenslauf" in H. Weber, ed., *Bernhard Riemann's gesammelte mathematische Werke und wissenschlaftlicher Nachlass*, 2nd ed. (Leipzig: Teubner, 1892), 553.

70. For example, draw a part of such a line, choose the middle point as corresponding to zero, settle on a centimeter as the unit of length and the right-hand direction as the positive direction to get

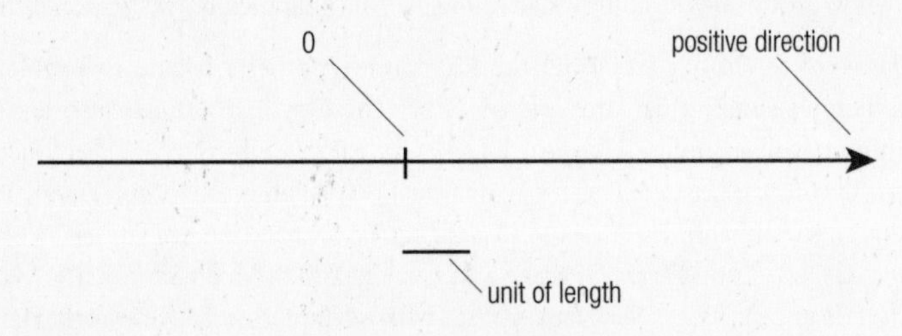

FIGURE 55. *Choice of 0, unit of length, and positive direction*

To the number 1, we associate the point exactly 1 centimeter to the right of zero. To the number 2.25, we associate the number 2.25 centimeters to the right of zero. To negative numbers we associate points to the *left* of zero: -1 is associated to the point exactly one centimeter to the left of 0, -2.25 to the point 2.25 units to the left of zero, and so on.

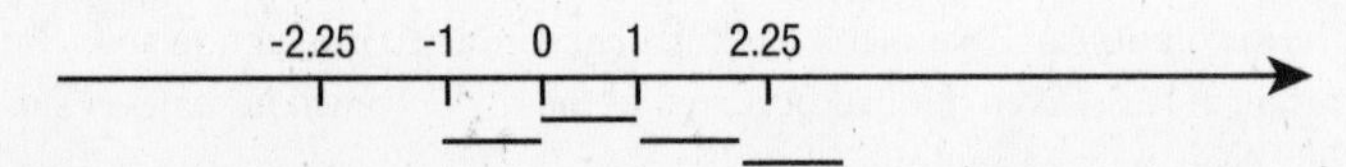

FIGURE 56. *The number line*

What about the number 100,000? It is not on this page. But it is out there, 100,000 centimeters to your right.

71. So, the pair (2, 3) corresponds to the point on the plane 2 unit lengths along the first number line in the positive direction and 3 along the second. The pair (–1, 2.2) corresponds to the point one unit to the left along the first number line and 2.2 units in the positive direction parallel to the second number line.

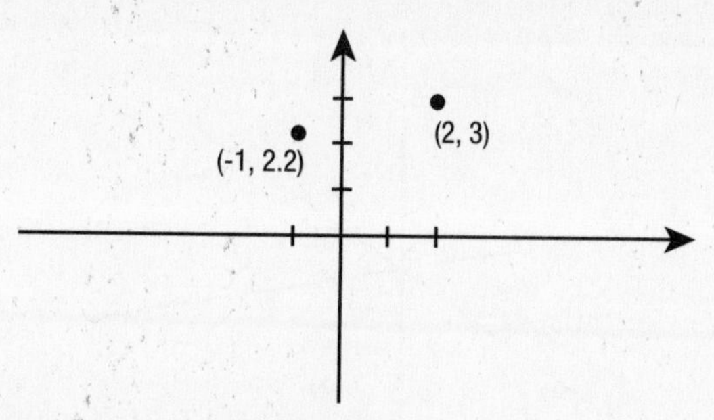

FIGURE 57. *The plane as pairs of real numbers*

Notice, by the way, that the order in which the numbers appear is important: We intend that (2, 3) and (3, 2) be different points! And they are. Had you oriented the lines differently, you would still be able to associate a pair of numbers with a point on a page in this book. Either way, you have a correspondence

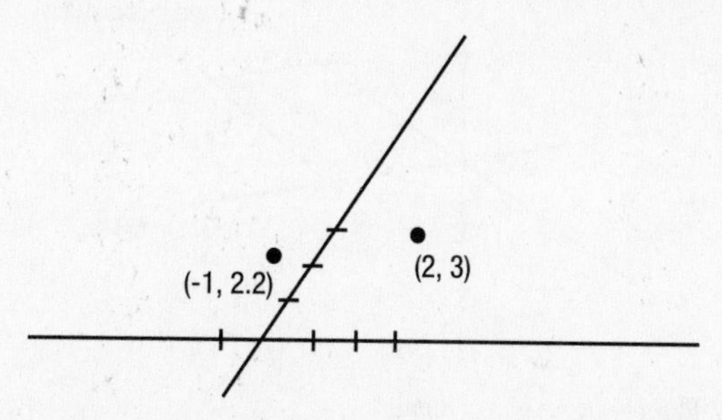

FIGURE 58. *Another correspondence between pairs of real numbers and the plane*

between pairs of numbers and points on an infinite page extending in all directions from the page of this book.

72. It is traditional to imagine these three lines tipped a bit, with a strong light behind them, and to draw their shadow on the plane, as in figure 59.

73. Adopting the convention of the note above, we would plot the triple (2, 3, –1) as in figure 60.

74. *Analysis* is the branch of mathematics that deals with the objects of calculus: functions, limits, rates of change, and what happens as things become vanishingly small and arbitrarily large. You can always think of a *function* as a machine that associates to the elements of one set, the elements of another. *Calculus* studies how functions change and provides a set of tools for determining the rate of

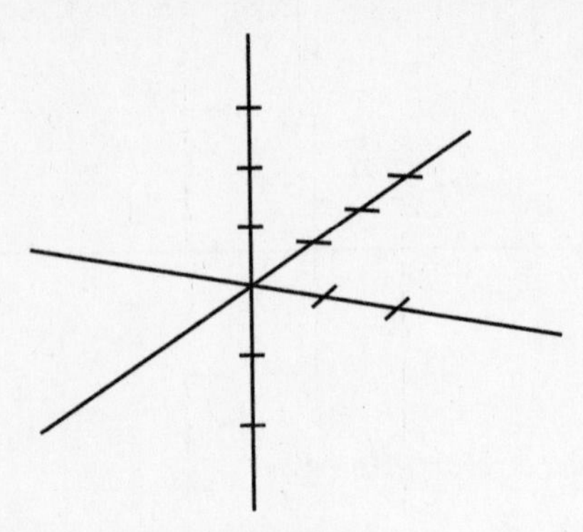

FIGURE 59. *Representing triples of real numbers*

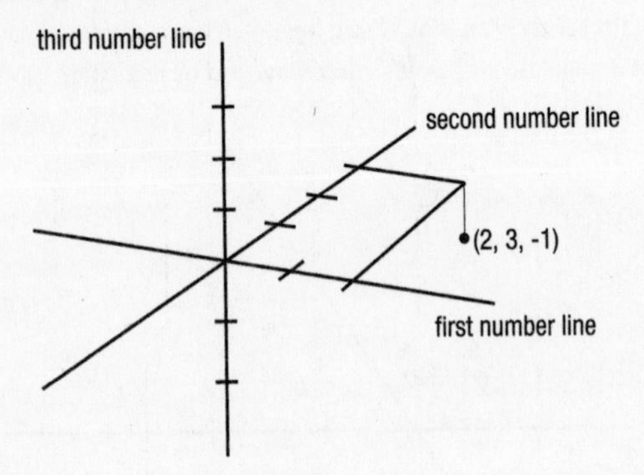

FIGURE 60. *Plotting triples of real numbers*

change of a function at some element of a set (this is called *differentiation*) and, inversely, for determining a function given its rate of change at every element of a set (this is called *integration*). Although the basic idea is simple, the details can be very complicated. One has to be quite careful defining rate of change (called *derivative*) of a function. This turns out to be another mathematical object, often another function, but one between different spaces, and care is needed, especially when the spaces are complicated. The entire beginning calculus course studies the special type of functions that associate real numbers to real numbers (that is, functions from **R**, or some interval in **R**, to **R**). The derivative (the rate of change) of any such function at a real number is a number.

75. For readers who remember some calculus, you can consider a function from **R** to a manifold as describing a path in the manifold: think of the real numbers as

time, and the value of the function at each time as the location at that time. If the manifold has dimension n, the derivative of such a function at any time turns out to be an n-tuple. Mathematicians distinguish between n-tuples representing derivatives and n-tuples representing locations in the manifold: The former are called *(velocity) vectors* and the latter *points*. One imagines the set of all vectors at a point of a manifold as the set of all velocities that curves that pass through that point can have. The set of vectors based at each point of an n-dimensional manifold are in one-to-one correspondence with $\mathbf{R}^n$ and so we think of each point of a n-dimensional manifold as having associated with it a space of velocity vectors that is a copy of $\mathbf{R}^n$. This is called the *tangent space* to the manifold at the given point, and is just the set of all possible velocities, or derivatives, of any curve through the point. To specify a metric, you have to specify the length of each velocity vector (which we think of as its speed) at each point. Riemann chose a particularly interesting class of metrics, by limiting himself to class of functions on the tangent space to a manifold at each point that were restrictive enough to be tractable, but general enough to capture a wide range of phenomena and be mathematically interesting.

76. That is, to define Euclidean 2-space, define the distance between two points (x_1, y_2) and (x_2, y_2) of $\mathbf{R}^2$ to be the square root of $(x_1 - x_2)^2 + (y_1 - y_2)^2$. To define $\mathbf{E}^3$, set the distance between two points (x_1, y_1, z_1) and (x_2, y_2, z_2) of $\mathbf{R}^3$ to be the square root of $(x_1 - x_2)^2 + (y_1 - y_2)^2 + (z_1 - z_2)^2$.

77. By definition, Euclidean n-space $\mathbf{E}^n$ is $\mathbf{R}^n$ with the distance between two points $(x_1, \ldots, x_n)$ and $(y_1, \ldots, y_n)$ defined to be the (positive) square root of $(x_1 - y_1)^2 + \cdots + (x_n - y_n)^2$. This is the generalization of the Pythagorean theorem to n dimensions.

78. Of course, this is far short of a proof (and you need a pretty big beach ball to see it clearly). One can construct a rigorous proof that geodesics are great circles from a symmetry argument. Alternatively, one can use calculus.

79. Since the Earth is nearly a sphere, the geodesics are very close to great circles. However, the Earth is not a perfect sphere: It is a bit flattened at each of the poles, and mountains and valleys change the curvature at different points. So the geodesics are not quite the same as great circles, but they are very close. This is because mountains, valleys, and flattening at the poles are relatively small compared to the size of the Earth.

80. Alternatively, think of the direction in which you want to go at a point as a little line tangent to the sphere at the point. Now, there is exactly one plane in $\mathbf{R}^3$ through any line and a point not on that line. Hence, there is exactly one plane through the center of the sphere, and containing the tangent line representing

the direction. This will intersect the sphere in a great circle in the desired direction. (It is a great circle since the plane goes through the center of the sphere.)

81. Riemann allows spheres of any dimension. In particular, the three-sphere that we worked so hard to define in chapter 3 is just homeomorphic to the set of points at fixed distance to some fixed point in $\mathbf{E}^4$.

82. Another notion that relies on the notion of homeomorphism, and which is often confused with homemorphism, is equivalence via *ambient homeomorphism.* When classifying surfaces topologically, we require only that homeomorphism map one surface to another. If you are looking at surfaces in three-space, one makes the (additional, stronger) requirement that there be a homeomorphism from three-space to itself that is a one-to-one correspondence between the surfaces in question. This is also the appropriate notion of equivalence for knot theory—a torus and one that is knotted are not necessarily equivalent.

83. The word *topology* was first used by Johann Benedikt Listing (1808–82) in his book *Vorstudien zur Topologie* (n.p.: 1847). An older expression, *analysis situs,* meaning "analysis of place" was used earlier, and in fact was more common than *topology* until the 1920s. In his book, Listing defines topology as "the study of the qualitative laws of relations between place" and expresses his strong belief that the nascent science was a worthy object of research and would yield deep results: "By topology, we understand then the study of the qualitative features of spatial forms, or the laws of connectivity, of the mutual position and of the order of points, lines, surfaces, bodies as well as their parts and their unions, abstracted from their connections with measure or size." Although he did not study topology for its own sake, but because of its applications to analysis, Riemann did a huge amount to advance topology.

84. For a history of the development of topology up to the work of Poincaré, see Jean-Claude Pont, *La topologie algébrique des origines à Poincaré* (Paris: Presses Universitaires de France, 1974).

85. The number that mathematicians actually use is called the *genus,* and is equivalent to the minimum number of cuts needed to leave a rectangle.

86. Although it is not difficult to see this (the fact that a one-to-one map that preserves distances is a homeomorphism), it is often difficult to establish the relations between different notions of equivalence.

87. This is a theorem that was first proved by Hilbert.

88. The dimension needs to be a positive integer or infinity. Riemann also allows for discrete spaces, but we shall not need these here.

89. W. K. Clifford, *Mathematical Papers*, ed. R. Tucker (London: Macmillan, 1882), 21–22. (Originally appeared as W. K. Clifford, "On the Space-Theory of Matter," *Cambridge Philosophical Society Proceedings* 2 [1876].)

90. The *tractrix* is the curve on the plane traced by the end of a string of fixed length that initially points straight up as the other end of the string moves in a straight line horizontally on the plane.

91. The story appeared as part of Fechner's collection *Vier Paradoxe* published in 1846 under the pseudonym of Dr. Mises. For more, see T. Banchoff's introduction to Abbott's *Flatland* cited in the note below.

92. For two recent editions, with notes and introductory essays by two geometers, see E. A. Abbott, *Flatland: A Romance of Many Dimensions*, introduction and notes by T. Banchoff (Princeton: Princeton University Press, 1991); and E. A. Abbott, *The Annotated Flatland: A Romance of Many Dimensions*, introduction and notes by I. Stewart (Cambridge, MA.: Perseus, 2002).

93. J. Collins, *Good to Great: Why Some Companies Make the Leap . . . and Others Don't* (HarperCollins, New York, 2001).

94. B. Riemann, *Gesammelte mathematische Werke und wissenschaftlicher Nachlass*, 3rd ed., von Heinrich Weber (Leipzig: B. G. Teubner, 1892), 541–58.

95. It is a myth that Riemann's work was all conceptual. Riemann's notebooks contain reams of computations.

96. Bismarck was much more than a hard-line conservative; he created the social security net of modern times.

97. It is not fair to blame Bismarck for the First World War. Kaiser Wilhelm II, who dismissed Bismarck shortly after acceding to power, deserves much more blame.

98. Even the disastrous Göttingen Seven incident worked to the university's favor. When the seven had been dismissed in 1837, they were replaced. Once hired back years later, the faculty size increased.

99. For more on Göttingen in the late nineteenth and early twentieth centuries, see C. Reid's biographies *Hilbert* (New York, Heidelberg, Berlin: Springer-Verlag, 1970) and *Courant in Göttingen and New York, The Story of an Improbable Mathematician* (New York, Heidelberg, Berlin: Springer-Verlag, 1976). Much information is contained in D. E. Rowe's articles "Klein, Hilbert and the Göttingen Mathematical Tradition," *Osiris* 5, no. 2 (1989): 186–213 and "'Jewish Mathematics' at Göttingen in the Era of Felix Klein," *Isis* 77 (1986): 422–49.

100. The German text was printed as a small pamphlet that was not easily available. However, it was reprinted, and Italian and French translations were prepared and appeared in major journals. An English translation by M. W. Haskell, "A Comparative Review of Recent Researches in Geometry," appeared in *Bulletin of the New York Mathematical Society* 2 (1893): 215–49. It has been reprinted in D. G. Saari (ed.), *The Way It Was: Mathematics from the Early Years of the Bulletin* (Providence: American Mathematical Society, 2003).

101. Examples are the integers with the operation of addition, homeomorphisms of a space with the operation of composition, permutations of the letters of a given word. *Group theory* is subfield of algebra that studies groups. One should not get the impression that every set with a single operation on its elements is a group. There are also other sets with a single operation that satisfy different axioms (semigroups, monoids, pseudogroups) that are of interest. One also studies sets with more than one operation. Depending on the axioms the two operations satisfy and the way they are linked with one another, one gets rings, fields, modules, lattices, and other structures. Maddeningly, the word *algebra* is also used to denote a particular class of ring (so *an algebra* is a particular example of a class of structures that algebra studies).

102. One needs pseudogroups and sheaves of groups instead of groups.

103. "Sur les functions fuchsiennes," *Comptes rendus de l'Academie des sciences* 92 (14 Feb 1881): 333–35: 92 (21 Feb 1881): 395–96; 92 (4 Apr 1881): 859–61.

104. Basic biographical details of Poincaré's life are readily available. There is, however, no good critical biography of Poincaré. For some interesting speculation on why this should be so, see Heinzmann, "Éléments preparatoires à une biographie d'Henri Poincaré." Heinzmann's article also contains an excellent sketch of Poincaré's childhood, with careful references.

105. Raymond Poincaré (1860–1934) was prime minister from 1912 to 1917, 1922 to 1924, and 1926 to 1929. He was president from 1913 to 1920.

106. Notebook by Poincaré's sister Aline, Item B 250, Documents sur Poincaré, Archive Henri-Poincaré, Université Nancy 2 (LPHS-AHP), p. 191. Cited in G. Heinzmann, "Éléments preparatoire à une biographie d'Henri Poincaré."

107. Ibid., p. 191.

108. Gaston Darboux, "Éloge Historique d'Henri Poincaré," read in a public lecture in December 1913, reprinted in *Oeuvres de Henri Poincaré,* vol 2 (Paris: Gauthiers-Villars, 1952).

109. There is an amusing divergence between Darboux's account of Poincaré's diligently learning German by going down to the café so that he could read German newspapers and tell his father and others what was going on, and his sister's account that Poincaré learned it because the German billeted with the family would sit in the warmest part of the drawing room. Unstated, but difficult to miss, is her somewhat tart assessment that Poincaré would have learned anything if it meant sharing the warm spot.

110. The *grandes écoles* are a system of 160 relatively well-funded small universities with carefully thought-out curricula. They have a very favorable student–teacher ratio (the 160 schools graduate about 11,000 students a year), modest tuition, and highly competitive admissions based on nationally administered written, and sometimes oral, exams. They supply almost 70 percent of the administrators and managers in French companies, and an even higher percentage of the top civil service.

111. Years later, Darboux, who had been chair of the examining committee for the thesis, would write, "From the first glance, it was clear to me that it was out of the ordinary and amply merited acceptance. It certainly contained enough results to furnish material for many good theses." (Darboux's eulogy, cited in note 101, 21.)

112. The university was rebuilt after the war, and inaugurated in 1957.

113. From Poincaré, "Science and Method," trans. by Francis Maitland. In *The Value of Science: Essential Writings of Henri Poincaré*, ed. Stephen Jay Gould (New York: The Modern Library, 2001), 392.

114. Ibid.

115. Ibid.

116. The three supplements to his essay for the Academy of Sciences in Paris (28 June 1880; 6 September 1880; 20 December 1880) outlined the fundamentals of non-Euclidean geometry and their relation to Fuchsian functions. These were forgotten and languished in the Archives of the Academic of Science in Paris, where they were rediscovered in December 1979 ninety-nine years later by Jeremy Gray, then a graduate student (and now a distinguished mathematical historian). For the texts of these supplements, and historical commentary on them, see eds. Jeremy J. Gray and Scott A. Walter, *Henri Poincaré, Three Supplementary Essays on the Discovery of Fuchsian Functions* (Berlin: Akademie Verlag GmbH and Paris: Albert Blanchard, 1997).

117. On April 11, the well-known Swedish mathematician, Mittag-Leffler, after having first written to Hermite, wrote Poincaré. They were to become lifelong

friends. Mittag-Leffler was planning to begin a new mathematical journal and invited Poincaré to publish in it.

118. The letters are reprinted in the collected works of Poincaré (*Oeuvres*, vol. 2) and in Klein's. Poincaré's collected work includes a long quote from *The Value of Science*, and a selection from Klein's account of the discovery (that differs slightly from the 1927 account). There seems to be no translation into English of the whole correspondence (and so I have translated the parts excerpted here and elsewhere).

119. I have translated the French verb *apercevoir* as "to glimpse" and *obtenir* as "to obtain." The two verbs have different connotations and Poincaré's choice of them is deliberate. Try switching them if you are not convinced.

120. [*die elliptischen Modulfunktionen*]

121. [*Kreisbogenpolygone*]

122. The motto is reputed to be an allusion to the defeat of Charles le Téméraire before Nancy in 1477. It is often translated colloquially in French as *"Qui s'y frotte, s'y pique,"* which translates roughly as "Meddle, and you'll get stung," or "Provoke me and you'll pay."

123. F. Klein, *Vorlesungen über die Entwicklung der Mathematik im 19 Jahrhundert*, Teil I, II. (Berlin: Springer, 1926 [Teil I], 1927 [Teil II]). See especially p. 249 of vol. 1: "We cannot conclude otherwise than that the Göttingen atmosphere saturated with geometric overtones exerted a strong compelling force on the impressionable and gifted Riemann. The surroundings in which a man finds himself are even more important that the facts and concrete knowledge offered him!" An English translation of the first volume, entitled *Development of Mathematics in the Nineteenth Century*, by M. Ackerman has appeared vol. 9 of the series Lie Groups: History, Frontiers and Applications, ed., R. Hermann (Brookline: Math Sci Press, 1979).

124. The complete letter follows. The third paragraph is one that Klein would never forget. Klein had sent Poincaré page proofs of an article announcing a big theorem, and Poincaré told Klein that he had known the result for some time. On receipt of Klein's note, Poincaré rushed an announcement off to *Comptes Rendus* (reference). Klein's bitterness is manifest in the extract that accompanies the letters in the French collected works. What is clear a century later is that both men "knew" the result to be true, but could not fully prove it. The mathematical tools were not then available, and the full proof would be completed by Poincaré and Koebe in 1910.

Paris, 4 April, 1882.

Je viens de recevoir votre lettre et je m'empresse de vous répondre. Vous me dites que vous désirez clore un débat stérile pour la Science et je ne puis que vous féliciter de votre résolution. Je sais qu'elle ne doit pas vous coûter beaucoup puisque dans votre note ajoutée à ma dernière lettre, c'est vous qui dites le dernier mot, mais je vous en sais gré cependant. Quant à moi, je n'ai ouvert ce débat et je n'y suis entré que pour dire une fois et une seule mon opinion qu'il m'était impossible de taire. Ce n'est pas moi qui le prolongerai, et je ne prendrais de nouveau la parole que si j'y étais forcé; d'ailleurs je ne vois pas trop ce qui pourrait m'y forcer.

Si j'ai donné votre nom aux functions kleinéennes, c'est pour les raisons que j'ai dites et non pas comme vous l'insinuez, *zur Entschädigung*; car je n'ai à vous dédommager de rien; je ne reconnaîtrai un droit de propriété antérieur au mien que quand vous m'aurez montré qu'on a avant moi étudié la discontinuité des groupes et l'uniformité des functions dans un cas tant soit peu général et qu'on a donné de ces functions des développements en series. Je réponds à une interrogation que je trouve en note à la fin d'une page de votre lettre. Parlant des functions définies par M. Fuchs au tome 89 de *Crelle*, vous dites; "Sind diese Funktionen wirklich eindeutig? Ich verstehe nur dass sie in jedem Wertsystem welches sie erreichen unverzweigt sind." Voici ma réponse, les functions étudiées par M. Fuchs se partagent en trois grandes classes; celles des deux premières sont effectivement uniformes; celles de la troisième ne sont en général que *unverzwiegt*. Elles ne sont uniformes que si l'on ajoute une condition à celles énoncées par M. Fuchs. Ces distinctions ne sont pas faites dans le premier travail de M. Fuchs; on les trouve dans deux notes additionelles, malheureusement trop concises et insérées l'une au *Journal de Borchardt*, t. 90, l'autre aux *Göttinger Nachrichten*, 1880.

Je vous remercie beaucoup de votre dernière note que vous avez eu la bonté de m'envoyer. Les résultats que vous énoncez m'intéressent beaucoup, voici pourquoi; je les avais trouvés il y a déjà quelques temps, mais les publier parce que je désirais éclaircir un peu la demonstration; c'est pourquoi je désirais connaître la vôtre quand vous l'aurez éclaircie de votre côté.

J'espère que la lutte, à armes courtoises, d'ailleurs, à laquelle nous venons de nous livrer à propos d'un nom, n'altérera pas nos bonnes relations. Dans tous les cas, ne vous en voulant nullement pour avoir pris l'offensive, j'espére que vous ne m'en voudrez pas non plus de m'être défendu. Il serait ridicule d'ailleurs, de nous disputer plus longtemps pour un nom, *Name ist Schall und Rauch* et après tout ça m'est égal, faites comme vous voudrez, je ferai comme je voudrai de mon côté.

Veuillez agréer, Monsieur, l'assurance de ma considération la plus distinguée,

Poincaré

125. If this strikes you as overstated, consider that Poincaré could, for instance, have used instead the still insulting but less loaded, "It cannot have been too difficult for you."

126. Both Poincaré and Klein would have been familiar with Goethe's *Faust*. The words in question come in response to Gretchen's "Gretchenfrage," asking what Faust thinks of religion. After long lines of evasion where Faust talks around the issue of what constitutes God or the divine, he finally states, *"Gefühl ist alles / Name ist Schall und Rauch,"* which translates as "Feeling is all, a name is but noise and smoke." Even today people use the latter half of the quote to downplay the significance of semantics, but Klein would also have instantly thought of the first part.

127. Klein's endnote to the Correspondence (p. 621 of vol. 3 of his collected works) confirms that this was the last correspondence between them. "Mit diesem Briefe fand die Korrespondenz seinerzeit ihr Ende. Ich vermochte es nur noch, die Abh. CIII fertigzustellen und mußte mich dann, wegen des Versagens meiner Gesundheit, von der weiteren Mitarbeit an der Theorie der automorphen Funktionen zurückziehen, wie schon oben auf S. 585 und in Bd. 2 dieser Ausgabe, S. 258 ausgeführt wurde. Auf die Übersendung meiner Arbeit habe ich von H. Poincaré keine Antwort mehr erhalten. Auch spätere persönliche Bezugnahme haben die hier berührten Fragen nur wenig geklärt."

128. From Felix Klein, *Vorlesungen uber die Entwicklung der Mathematik im 19 Jahrhundert*, Teil I (Berlin: Springer, 1928). There is an English translation, *Development of Mathematics in the Nineteenth Century* by M. Ackermann (Brookline: Math Sci Press, 1979). The passage quoted can be found on p. 361: "The price I had to pay for my work was extraordinarily high—my health completely collapsed. In the next years I had to take long leaves and renounce all productive activity. Things did not go well again until the autumn of 1884, but I have never regained my earlier level of productivity. I never returned to elaborate my earlier ideas. And later, when I was at Göttingen, I returned to extending the domain of my work and to general tasks of organizing our science. Thus one can understand that since then I have only occasionally touched on automorphic functions. My real productive work in theoretical mathematics perished in 1882. All that has followed, insofar as it has not been purely expository, has been merely a matter of working out details."

129. He married on April 20, 1881. His spouse was Mlle. Poulain d'Andecy, of the family of Geoffroy Saint-Hilaire.

130. Maître de conferences, Poincaré file, Centre historique des Archives nationales, Paris, AJ/16/6124.

131. There is a useful timeline in the online archives maintained by the Université de Nancy (www.univ-nancy2.fr/poincare/index.html). The National Archives contain appointment dates. The archives of the Academy of Science show that Poincaré was first nominated in 1880 and a number of times thereafter. But there was quite a backlog.

132. Recall that by Riemann's work specifying a geometry boils down to deciding how to measure the length of a velocity vector (that is, the speed) at a point of a curve traced by an object on the torus. We can agree to think of a curve on the torus as a curve traced on the square, and we are agreeing to use the standard Euclidean metric on the latter. We will not run into any trouble because angles match up when we make identifications—in particular, four right angles fit together round a vertex. In particular, for readers with some calculus, we are agreeing to define the length of a velocity vector at each point of an object moving along a path on the torus, by considering the corresponding path on the square and taking the Euclidean length of the velocity of the corresponding motion of the object.

133. Note that you also have straight lines that are finite in length. Such lines go around the torus a fixed number of times in one direction and a different number of times in the other. There are also straight lines that are infinite in length: They go around the torus an infinite number of times in both directions and come arbitrarily close to any point on the torus, a situation that is described by saying that the line is *dense* on the torus.

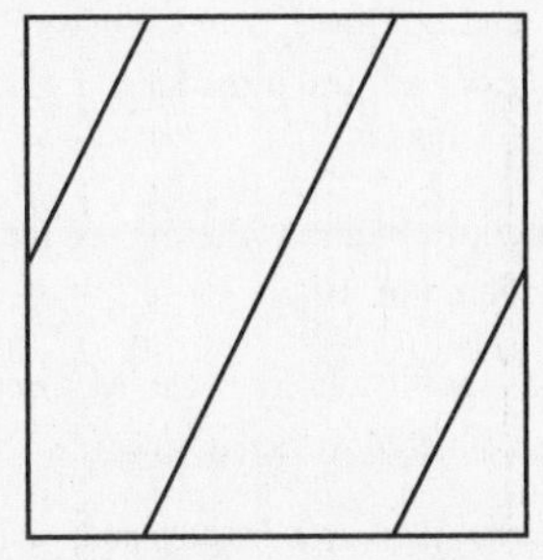

FIGURE 61. *A finite straight line on the torus (opposite edges connected)*

134. In fact if the geometry is Euclidean, so the sum of the angles of a triangle is 180 degrees, the sum of the angles in any octagon, regular or not, has to be 1,040 degrees. The reason is that if you pick one vertex of the octagon and draw straight lines from it to each other vertex, you divide the octagon into 6 triangles whose angle sums equal the sum of the angles of the octagon. Thus the octagon has 180 times 6 degrees, or 1,040 degrees. (This also allows us to conclude that the degree measure of each angle of a regular octagon is 1040 divided by 8, which equals 135.)

135. One needs to know that an n-holed torus can be represented as a $4n$-gon with alternating pairs of edges identified in the same pattern as for a two-holed torus. For a nice account for readers with modest mathematical backgrounds see chapter 11 of J. Weeks, *The Shape of Space*, 2nd ed. (New York, Basel: Marcel Dekker, 2002). For more mathematically demanding treatments see J. Stillwell, *Geometry of Surfaces* (New York: Springer-Verlag, 1992) or A. Beardon, *The Geometry of the Discrete Groups* (New York: Springer-Verlag, 1983).

136. In fact, if you consider the torus as the unit circle $\mathbf{S}^1$ in $\mathbf{E}^2$ and consider the torus $\mathbf{S}^1 \times \mathbf{S}^1$ as a subset of $\mathbf{E}^2 \times \mathbf{E}^2$ (which is the same as $\mathbf{E}^4$ because a pair of pairs of real numbers is a quadruple of real numbers, and vice versa), then this torus inherits a distance from the Euclidean distance in $\mathbf{E}^4$. If you enjoy computation, you can show that with this distance, this torus is flat by checking that arbitrary triangles have 180 degrees. Incidentally, higher dimensional tori $\mathbf{S}^1 \times \ldots \times \mathbf{S}^1$ also have a natural flat metric, and the same argument works.

137. The proof of the Riemann embedding theorem appears in John Nash, "The Imbedding Problem for Riemannian Manifolds," *Annals of Mathematics* 63 (1956): 20–63. The biography cited in the text is Sylvia Nasar, *A Beautiful Mind* (New York: Simon and Schuster, 1998) and the movie has the same title.

138. Poincaré, "Science and Hypothesis," *The Value of Science: Essential Writings of Henri Poincaré*, ed. Stephen Jay Gould (New York: The Modern Library 2001), p. 56.

139. For a careful account, see D. Goroff's introductory essay in H. Poincaré, *New Methods of Celestial Mechanics*, ed. and introduced by D. Goroff (New York: American Institute of Physics, 1993).

140. For a good, popular account of chaos theory, see James Gleick, *Chaos: Making a New Science* (New York: Penguin, 1988).

141. This retrospective was published in *Acta Mathematica*, vol. 38, 1921, and reprinted in Poincaré's collected works. (*Oeuvres*, 2: 183.)

142. "Analysis Situs" in *Journal de l'École Polytechnique* 1 (1895): 1–121. (*Oeuvres*, 2: 193–288.) Analysis Situs is the then-common, but now old-fashioned, name for

topology. It translates roughly as "the analysis of place" and was first used by Leibniz.

143. The Betti numbers were presumed to be topologically invariant, but the lack of proof was to become an embarrassment. The fact that the Betti numbers were invariant under homeomorphism had to await James Alexander's work. He proved it in dimension three in 1913. The idea behind Alexander's proof worked in general, and Alexander and his supervisor, O. Veblen, both of whom we shall meet later, proved it in general a few years later.

144. Ibid. (*Oeuvres*, 6:258.)

145. "Complément à l'analysis situs," *Rendiconti del Circolo Matematico di Palermo* 13 (1899): 285–343. "Second complément à l'analysis situs," *Proceedings of the London Mathematical Society* 32 (1900): 277–308. "Certaines surfaces algébriques; troisième complément à l'analysis situs," *Bulletin de la Société Mathématique de France* 30 (1902): 49–70. "Sur les cycles des surfaces algébriques; quatrième complément à l'analysis situs," *Journal de Mathématiques* 8 (1902): 169–214 (Liouville's journal). "Cinquième complément à l'analysis situs," *Rendiconti del Circolo Matematico di Palermo* 18 (1904): 45–110. All, together with the research announcements, are reprinted in vol. 6 of *Oeuvres*.

146. The third complement studies the particular class of algebraic surfaces of the form $z^2 = F(x,y)$.

147. *Oeuvres*, 6:238.

148. *Oeuvres*, 6:270.

149. *Oeuvres*, 6:435.

150. Poincaré certainly didn't know that the space he had discovered could be described by identifying opposite faces of a dodecahedron. That description appeared in C. Weber and H. Seifert, "Die beiden Dodekaedräume," *Mathematische Zeitschrift* 37, no. 2 (1933): 237. For a really nice, readable description of the two dodecahedral spaces in this article see Jeff Weeks, *The Shape of Space: How to Visualize Surfaces and Three-Dimensional Manifolds* (New York: Marcel Dekker, 1985). A second edition appeared in 2002. I would recommend this book unreservedly. It can be read by nonmathematicians and, in particular, by an interested high school student. Another more geometric way to describe the Poincaré dodecahedral space is to consider the space of all dodecahedra that can be inscribed in a two-dimensional sphere of radius one. That is, we think of each such dodecahedron as a point. This is analogous to thinking of three-space as the set of all triples of real numbers. The space of all dodecahedra in the unit sphere is

three-dimensional, because it takes two parameters to specify one vertex on the two-sphere, and a third to describe the direction in which you must head to get to the next. In other words, specifying three numbers specifies a dodecahedron. To get a curve that cannot be shrunk to a point, consider the set of dodecahedra with a fixed vertex. This corresponds to traversing all angles between zero and 120° from the fixed vertex: The dodecahedra obtained at the beginning (zero degrees) and end (120°) are the same. See also J. Milnor, "The Poincaré Conjecture One Hundred Years Later" (www.math.sunysb.edu/~jack) and "Towards the Poincaré Conjecture and the Classification of Three Manifolds," *Notices of the American Mathematical Society* 50, no. 10 (2003): 1226–33. Milnor's description is as the space of regular icosahedra inscribed in the two-dimensional sphere.

151. *Oeuvres*, 6:498.

152. The number of Nobel nominations comes from J. Mawhin, "Henri Poincaré. A Life in the Service of Science," *Notices of the American Mathematical Society* 52 (2005): 1036–44. This is the text of a lecture presented at the Poincaré Symposium in Brussels, Oct 8–9, 2004, on the occasion of the 150th anniversary of Poincaré's birth.

153. *Science and Hypothesis* is a translation of *La Science et l'hypothèse* which appeared in 1902. A revised and corrected edition appeared in 1906 and is still in print. *La Valeur de la science* (translated in 1958 as *The Value of Science*) appeared in 1905. The third volume *Science et Méthode* (translated as *Science and Method*) of the trilogy appeared in 1908. An English translation of the trilogy is available under the title *The Value of Science: Essential Writings of Henri Poincaré* (New York: Random House, 2001). A posthumous book *Dernières Pensées* appeared in 1913 (translated into English as *Mathematics and Science: Last Essays* in 1963).

154. The Institut de France was founded in 1795 after the suppression two years earlier in the French Revolution of the learned societies formed in the mid-seventeenth century. The Institut consists of five sections, the French Academy (40 members, language and literature, founded in 1635 by Cardinal Richelieu), the Academy of Fine Arts (55 members, formed in 1816 by joining the Academy of Painting and Sculpture, founded in 1648, and the Academy of Music, founded 1671), the Academy of Inscriptions and Belles Lettres (55 members, history and archaeology, founded 1663), the Academy of Science (190 members, medicine, mathematics and science, founded 1666), and the Academy of Ethics and Political Sciences (50 members, founded 1795, suppressed 1803, reestablished 1832).

155. Poincaré, Introduction, *The Value of Science*, trans. G. B. Halstead, 189–90.

156. Dehn began as student of Hilbert in Göttingen in 1899. In 1900, Hilbert had presented a list of problems at a conference in Paris. The list became famous almost overnight.

157. You can compute the areas of polygons in the plane by dividing them up into finitely many triangles and adding up all their areas. If you are careful, you can even build a definition of area for polygonal plane figures this way. Dehn's work showed that any such approach in three-space was doomed. Even for the most basic polyhedral objects, you were going to have to resort to adding the volumes of infinitely many objects and, hence, to calculus. Dehn's proof was extremely elegant. He found an invariant of indecomposability that allowed him to recast the problem as a problem in elementary number theory that he could solve. (For a careful development of Dehn's invariant, see Dupont and Sah, "Scissors Congruences," *Journal of Pure and Applied Algebra* 25 (1982): 159–95.) Dehn's invariant actually plays a role in three-manifold theory today. Using the fact that three-manifolds can be defined by identifying faces of a polyhedron, Thurston tweaked Dehn's invariant to give a geometric invariant of the manifold. Mostow rigidity then implies that geometric invariants are topological invariants.

158. Exactly when Dehn met Heegaard is unclear. See J. Stillwell, "Max Dehn," 965–78, in *History of Topology*, ed. I. M. James (Amsterdam: Elsevier, 1999). According to Stillwell (p. 968), they met in 1903 or 1904. Johannson's obituary of Heegaard says they met in Kassel in 1903, but Dehn's widow Toni told Magnus they first met at the Heidleberg ICM in 1904.

159. M. Dehn and P. Heegaard, "Analysis Situs," in *Enzyklopädie der Mathematischen Wissenschaften III* AB 3 (Leipzig: Teubner, 1907), 153–220. This article was written in 1905 and contained a rigorous proof of the the classification theorem for compact surfaces that had been discovered by Möbius in the orientable case and by Dyck in general. A footnote at beginning states, "Heegaard undertook the collection of literature for the the article, as well as working out the essential portions. Responsibility for the final form of the article is Dehn's."

160. Moritz Epple shows that Dehn tried to prove the Poincaré conjecture in 1908 and thought that he had succeeded (Stillwell, 969). Tietze found a mistake in Dehn's proof and the paper was withdrawn. As Volkert says, Dehn nearly became "the first victim of Poincaré's conjecture." (See K. Volkert, "The Early History of Poincaré's conjecture," in *Henri Poincaré, Science and Philosophy*, eds. J. L. Greffe, G. Heinzmann, and K. Lorenz [Berlin: Akademie-Verlag and Paris: Albert Blanchard, 1996], 241–50.)

161. M. Dehn, "Über die Topologie des dreidimensionale Raumes," *Mathematische Annalen* 69 (1910): 137–68.

162. More precisely. Dehn shows that the fundamental groups of the homology spheres act on the hyperbolic plane in a canonical way.

163. Dehn's lemma states that if a closed curve in the three-sphere bounds a piecewise linear disk in such a way that an annulus along the boundary is free of sings, then the curve actually bounds a regularly embedded disk. A correct proof was not found until 1957 by Greek mathematician, C. D. Papakyriakopoulos ("On Dehn's lemma and the asphericity of knots," *Proceedings of the National Academy of Sciences U.S.A.* 43 [1957]: 169–72 and *Annals of Mathematics* 66 [1957]: 1–26). The lemma was used to introduce a new criterion for knottedness (namely, a closed curve is unknotted only if the fundamental group of its complement is abelian). Dehn's lemma allows you to interpret relations in the fundamental group geometrically in terms of disks bounding the loops representing the elements.

164. H. Tietze, "Über die topologischen Invarianten mehrdimensional Mannigfaltigkeiten," *Monatshefte für Mathematik und Physik* 19 (1908): 1–118. This paper was critical in disseminating Poincaré's topological ideas. Tietze underscored the importance of the fundamental group in deriving invariants of manifolds, and had realized that some of Wirtinger's ideas allowed one to compute the fundamental groups of knot complements. He pointed out a number of problems with Poincaré's definition of homological invariants. He showed that the notion of knot equivalence needed attention to avoid wild knots. Wild knots, for instance, need not bound a disk in the usual sense of the word. Tietze first asked the question about whether the complements of two knots could be homeomorphic without the knots being isotopic to each other or their mirror image (§15). This became a famous question that was only recently solved. He asked whether all submanifolds of Euclidean three-space bounded by a torus were knot complements. He looked at complements of unions of left- and right-trefoils, pointing out that no one had shown that left- and right-hand trefoils were not equivalent. In analogy with Riemann's method of studying surfaces as surfaces covering the two sphere but branched over finitely many points, he studied three manifolds that covered the three-sphere but that were branched over links. He asked whether all three-manifolds could be so-obtained (§18).

165. J. Hadamard, "L'oeuvre mathématique de Poincaré," *Acta Mathematica* 38 (1921): 203–87. This is the definitive retrospective of Poincaré's work, written by the leading French mathematician of the years immediately succeeding Poincaré.

For a very beautiful account of topology in the first half of the twentieth century, see C. McA. Gordon, "3-Dimensional Topology up to 1960," 449–89 in *History of Topology*, ed. J. M. James (Amsterdam: Elsevier, 1999).

166. For a good account of the relationship between the two men, see P. Galison, *Einstein's Clocks, Poincaré's Maps: Empires of Time* (New York: W. W. Norton, 2003).

167. H. Poincaré, "La mesure du Temps," *Revue de metaphysique et de morale* 6 (1898): 371–84; "Sur la dynamique de l'électron," *Rendiconti del Circolo Matematico di Palermo* 21 (1906): 129–75 (announcement in *Comptes rendus de l'Academie des sciences* 140 (1905): 1504–08).

168. P. Galison, *Einstein's Clocks, Poincaré's Maps: Empires of Time* (New York: W. W. Norton, 2003).

169. Klein's American students Cole and Fine from his Leipzig days became heads at the University of Michigan and Princeton, respectively. From Göttingen, Haskell went to the University of Michigan, then Berkeley. Osgood and Bôcher were the source of strength of the Harvard mathematics department. Two of Klein's German students, Oskar Bolza and Heinrich Maschke from Leipzig, emigrated to the United States and were the mainstays of the mathematics department of the University of Chicago. Van Vleck, who was also a student of Klein's at Göttingen, was a leading figure in establishing the mathematics department at the University of Wisconsin.

170. Any time you have an expression that is invariant under transformations, then any multiple of that expression is also invariant. Hence, there are always infinitely many invariants for a given set of equations and a given group of transformations. More is true, however. The sum and product of two invariants is again an invariant. Thus the set of invariants is closed under multiplication by complex numbers and you can add and multiply its elements. This is a structure that mathematicians call an *algebra*.

171. The details of this story are well known to all mathematicians. For a good account of Hilbert and this discovery, see Constance Reid, *Hilbert* (New York: Springer-Verlag, 1970).

172. Hilbert studied the set of relations between invariants, which had a different type of algebraic structure than the invariants (they had the structure of a polynomial module instead of the structure of an algebra). Then he studied the set of relations on the relations, and the set of relations on the set of relations on the set of relations, and so on. He showed that this process terminated, a result called the *Hilbert Syzygy theorem*.

173. D. Hilbert, *Die Grundlagen der Geometrie* (Leipzig: Teubner, 1899). English translation by E. J. Townsend (n.p.: 1902). Subsequent editions have been in print contiuously since 1902.

174. H. Poincaré, Review of Hilbert's "Foundations of Geometry," *Bulletin of the American Mathematical Society* 10 (1903): 1–23. Reprinted in D. Saari, ed., *The Way It Was: Mathematics from the Early Years of the Bulletin* (Providence: American Mathematical Society, 2003, 273–96).

175. In early 1914, Göttingen had over eight hundred mathematics students, over a hundred of whom were advanced. For an account, see C. Reid's books and the article of Rowe.

176. Letter of Poincaré to Giovanni Battista Guccia (December 9, 1911). Archives Circolo Matematico del Palermo. [Mon cher ami, Je vous ai parlé, lors de votre dernière visite, d'un travail qui me retient depuis deux ans. Je ne suis pas plus avancé et je me decide à l'abandoner provisoirement pour lui donner le temps de mûrir. Cela serait bien si j'étais sûr de pouvoir le reprendre; à mon âge je ne puis en répondre . . . Dites moi, je vous prie, ce que vous pensez de cette question et ce que vous me conseillez.] Letter of Guccia to Poincaré (December 12, 1911). Collection particulère Paris, image on Poincaré online archive, Nancy. [Mon cher ami, Je vous confirme ma dépêche: "Conseil publiez." Quoique inachevé, votre travail ouvrira certainement des voies nouvelles aux autres chercheurs, et la Science en profitera. Au surplus, si vous le croyez nécessaire, vous pourriez ajouter au commencement (sous forme de lettre ou dans une note), que c'est sur les instances prieés de la Directions des *Rendiconti* que vous vous étes decide à publier ces recherches inachevées.] The paper "Sur un théorème de géometrie" appeared in *Rendiconti del Circolo matematico di Palermo* 33 (1912): 375–407 and is reprinted in *Oeuvres*, 6:499–538.

177. Symplectic geometry is a rather different type of geometry that is built not on distances and angles (that is, a metric), but on area; and symplectic topology is the study of manifolds that can carry a symplectic structure. These fields have become exceedingly important in the last two and half decades.

178. P. Painlevé writing in *Le Temps*: "Henri Poincaré était vraiment le cerveau vivant des sciences rationelles." This is widely quoted.

179. H. Poincaré, "The present and future of mathematical physics," *Bulletin of the American Mathematical Society*, 12 (1906): 240–260; reprinted 37, no. 1 (1999):25–38.

180. There were 37 in the 1880s, 35 in the 1890s, 31 from 1900 to 1910. Wikipedia (en.wikipedia.org/wiki/World's_Fair). One of the largest displays at the Chicago World Fair was the Krupps pavilion.

181. Sadly, however, Palermo's blossoming built less on deep roots than on a temporary bubble inflated by exports of Sicilian goods that were inexpensive because of cheap labor. Things were out of control and civil society had actually broken down. Civil and governmental authorities though nominally in charge, had to call on the Mafia to put down labor revolts. Joseph Petrosino, a high-profile New York policeman who had traveled to Sicily tracing the roots of protection and extortion operations in the United States, was openly gunned down in 1909 under the Garibaldi's statue in the Piazza Marina in central Palermo.

182. A. Brigaglia: "The Circolo Matematico di Palermo," 179–200 in K. H. Parshall, A. C. Rice, eds., *Mathematics Unbound: The Evolution of an International Mathematical Research Communtiy, 1800–1945* (Providence: American Mathematical Society and London: London Mathematical Society, 2002). See p. 192.

183. In an address to a group of students, Poincaré had warned: "Do you think that Wilhelm II would have the same aspirations as you? Do you count on him using his power to defend your ideal? Or do you put your confidence in the people, and hope that they will commune in the same ideal? That's what one hoped in 1869. Do not imagine that what the Germans call right or liberty is the same thing that we call by the same names." H. Poincaré, "Le Banquet du 11 Mai," Annual Banquet of the General Association of Students of Paris, 1903, 63, quoted in Galison, 213.

184. "When the Great War broke out, it came to me not as a superlative tragedy, but as an interruption of the most exasperating kind to my personal plans." These are the words with which Vera Brittain opens her book, *Testament of Youth: An Autobiographical Study of the Years 1900–1925* (New York: Penguin, 1989; first published in London, 1933).

185. Einstein had submitted his *Habilitationschrift* and obtained an entry-level position at the University of Bern in 1908. He was recognized as a leading thinker by 1909 and able to resign his job at the patent office. He became a full professor at Karl-Ferdinand University in Prague in 1911, and returned to Switzerland in 1912, having obtained a chair at the celebrated Eidgenössische Technische Hochschule in Zurich. The offer he received from Berlin was very generous. He held a research position at the Prussian Academy of Sciences; had a chair, but no teaching duties, at the University of Berlin; and was offered the directorship of the nascent Kaiser Wilhelm Institute of Physics.

186. The quote is from the St Andrews Web site. Einstein presented two papers, "On General Relativity Theory" to the Berlin Academy on November 11 and 25; and Hilbert, a note "On the Foundations of Physics" to the Göttingen Academy on

November 20, 1915. Hilbert's paper was the first to contain the correct field equation of general relativity. The encounters between the two men were friendly, and Hilbert freely admitted that the idea was Einstein's. Hilbert's derivation was much more straightforward.

187. Heegaard's thesis: P. Heegaard, "Forstudier til en topologisk teori for de algebraiske fladers sammenhæng" (Kobenhavn: *Der Nordiske Forlag,* 1898). Alexander's translation into French appeared as: "Sur l'Analysis situs," *Bulletin de la Société Mathématique de France* 44 (1916): 161–242. For a biography of Alexander, see S. Lefschetz, "James Waddell Alexander (1888–1971)," *Yearbook of the American Philosophical Society (1973)* (Philadelphia: American Philosophical Society 1974), 110–14.

188. J. W. Alexander, "Note on Two 3-Dimensional Manifolds with the Same Group," *Transactions of the American Mathematical Society* 20 (1919): 339–42. Recall that Tietze had shown that two three-manifolds with the same fundamental groups necessarily had the same homology groups (this is definitely false if the manifold had dimension greater than three). The two manifolds that had the same fundamental groups are known nowadays as the lens spaces L(5,2) and L(5,1). For any positive integers p and any integer q less than p and greater than or equal to 0 and no divisor in common with p, you define the lens space $L(p, q)$ as follows: Divide the equator of the three ball into p equal segments, and think of the upper and lower hemispheres as two p-sided polygons. Connect these polygons by rotating the upper one q/pth a full turn counterclockwise. Tietze had noticed that $L(p, q)$ is covered by the three-sphere p times, so has finite fundamental group with p elements (the pth power of at least one of which is the identity). He had asked specifically whether L(5,1) and L(5,2), which are the simplest examples that are not obviously homeomorphic, were in fact homeomorphic. Alexander seems to have been unaware that Tietze explicitly asked this question. The term "lens space" comes from this paper: W. Threlfall and H. Seifert, "Topologische Untersuchung der Discontuitätsbereiche endlicher Bewegungsgruppen der dreidimensionalen sphärische Raumes I," *Mathematische Annalen* 104 (1930): 1–70, which thoroughly investigates such spaces. After dividing the equator of the three-ball into p pieces, the authors picture the upper and lower hemispheres as p-sided prisms, so that the three-ball looks like a $2p$-sided jewel.

189. He clarified the relation between combinatorial topology, in which three-manifolds are regarded as identifying polyhedra, and general topology in which three-manifolds cannot necessarily be cut up into a finite number of pieces with finitely many sides. He discovered, for example, the famous Alexander "horned sphere," which is homeomorphic to a two-dimensional sphere and lives in the

three-sphere, but does not divide it into two regions whose closures are homeomorphic to three-balls. On the other hand, he showed that a polyhedral two-sphere divides the three-sphere into two balls, the closure of each of which is a three-ball. He used a similar argument to show that any polyhedral torus in the three-sphere bounds a solid two-torus, a fact assumed without proof by Dehn and conjectured by Tietze (who pointed out that it was not obvious and needed proof).

190. J. W. Alexander, "Some Problems in Topology," *Verhandlungen des Internationalen Mathematiker Kongresses Zürich* (1932), Kraus Reprint (1967): 249–57.

191. He claimed that every open three-manifold that can be continuously deformed to a point is homeomorphic to $\mathbf{R}^3$.

192. Veblen's book in 1922 (O. Veblen, *Analysis Situs* [New York: American Mathematical Society, 1922]) discusses the Poincaré conjecture (chapter 5, §39, p. 147), but was not used as a text.

193. For more on Seifert's life, see the short biography, Dieter Puppe, "Herbert Seifert: May 27, 1907–October 1, 1996," in *History of Topology*, ed. I. M. James, (1021–27). The extract from Threlfall's diary is quoted therein. The German orignal reads, "Das Buch ist aus Vorlesungen hervorgegangen, die er eine von uns dem anderen im Jahre 1927 an der Technische Hochschule Dresden gehalten hat. Bald hat aber der Hörer so wesentlich neue Gedanken zur Ausarbeitung beigetragen und sie so von Grund auf umgestaltet, daβ eher als sein Name der des ursprünglichen Verfassers auf dem Titelblatte fehlen dürfte." *Lehrbuch der Topologie* appeared in 1934. An English translation appeared in 1980. The quotations regarding the Poincaré conjecture appear on p. 225 of the English translation. The English translation appears with a translation of Seifert's fundamental paper on fibred manifolds.

194. This is not the place to attempt a full history of mathematics in the United States. Probably the ablest American-born mathematician of the nineteenth century was Willard Josiah Gibbs (1839–1903). He received the first Ph.D. offered by Yale in engineering in 1863. But his mathematical development depended crucially on training he received in Paris, Berlin, and Heidelberg from 1866 to 1869.

195. Yale was the first to establish a graduate program in 1847, with the first three Ph.D.s (in philosophy and psychology, in classics, and in physics) awarded in 1861. Pennsylvania, Harvard, Michigan, and Princeton followed suit a little over a decade later. Johns Hopkins University, which focused explicitly on graduate education, opened in 1876 and really changed the landscape. But its brief success faltered when the university was unable to secure Klein to succeed Sylvester.

Clark University was another institution set up to focus especially on graduate programs. It had some real promise, but never fully realized it because it was underendowed. With the arguable exception of Harvard, which, thanks to Osgood and Bôcher, had real strength in analysis, there was no really first-rate graduate mathematics department in the United States. For an account from the perspective of mathematics, see K. H. Parshall and D. E. Rowe, *The Emergence of the American Mathematical Community 1876–1900: J. J. Sylvester, Felix Klein, and E. H. Moore* (Providence: American Mathematical Society, 1994). Their chapter 6 has a particularly good account, with copious archival references, of the mathematical landscape at the time and the founding of the University of Chicago.

196. Osgood received his doctorate from Erlangen, and Bôcher from Göttingen.

197. Four students of Moore's, L. E. Dickson (1874–1954), Oswald Veblen (1880–1960), R. L. Moore (1882–1974), and G. D. Birkhoff (1884–1944) were off the charts by any measure, and their effect on American mathematics is nearly incalculable. Each was an excellent mathematician, each had genuine administrative ability, and each was a superb graduate-level instructor. Birkhoff pushed the mathematics department at Harvard to true excellence; Veblen did the same at Princeton; R. L. Moore, at the University of Texas at Austin; and Dickson continued the tradition at the University of Chicago.

198. O. Veblen, "Theory on plane curves in non-metrical analysis situs," *Transactions of the American Mathematical Society*, 6 (1905) 83–98.

199. See L. B. Feffer, "Oswald Veblen and the Capitalization of American Mathematics: Raising Money for Research, 1923–1928," *Isis* 89 (1998): 474–97.

200. See, for example, G. D. Birkhoff, "Fifty Years of American Mathematics," *Science* 88, no. 2290 (1938): 461–67 (especially, 465).

201. They were James Alexander (1915–51, student of Veblen), Marston Morse (1892–1977, student of Birkhoff), Hassler Whitney (1907–89, student of Birkhoff), Solomon Lefschetz (1884–1972, student of William Story who was a student of Klein), and Norman Steenrod (1910–71, student of Lefschetz).

202. For a sense of Lefschetz's ebullience, even as an older man, see his essay, "Reminiscences of a Mathematical Immigrant in the United States," *American Mathematical, Monthly* 77 (1970): 344–50, written two years before his death.

203. In the first complement, Poincaré introduces dual cells and a "reciprocal" cell complex, together with a join operation on simplices that he asserts makes sense. It is difficult to make sense of these arguments without introducing cohomology.

204. Quoted in C. Reid, *Hilbert* (New York: Springer, 1970), 205.

205. Milnor showed that the total curvature of a knot that was not equivalent to the unknotted circle was greater than 4π. The result appeared in the Annals: J. Milnor, "On the Total Curvature of Knots," *Annals of Mathematics* 52 (1950): 248–57.

206. Two differentiable structures on homeomorphic manifolds are equivalent if there is a homeomorphism between the manifolds that is differentiable in both directions (that is, which is such that it, and its inverse, have well-defined derivatives at every point).

207. J. Milnor, "On Manifolds Homeomorphic to the 7-Sphere," *Annals of Mathematics* 64 (1956): 399–405.

208. For example, Milnor shows that the manifolds he is considering are spheres by producing a function on them that has only two critical points, that is, two points where the rate of change is zero in all directions. It had been known for a very long time that any continuous function on a bounded set has a maximum and a minimum. If, in addition, the function is differentiable, then at the maximum and minimum must be critical points of the functions: points at which the rate of change of the function in all directions is zero. What is deeper and less obvious is that if you have a function on a compact manifold that has derivatives everywhere and if it only has two critical points, then the manifold must be a sphere. The fact seemed to be a curiosity. In Milnor's hands, it became a central tool.

209. M. Kervaire, "A Manifold Which Does Not Admit and Differentiable Structure," *Commentarii Mathematici Helvetici* 34 (1960): 257–70. Kervaire's example is obtained from an exotic nine-dimensional sphere.

210. A critical point of a function is a point where all its derivatives are equal to zero.

211. J. Milnor, *Singular Points of Complex Hypersurfaces* (Princeton: Princeton University Press, 1968).

212. R. Thom, *Mathematical Models of Morphogenesis*, English trans. W. M. Brookes (New York: Halsted Press, 1983).

213. For instance, there was an updated version of Dehn's approach whereby you could consider a cube with a knotted hole. The exterior of this in the three-sphere is a solid torus. Perhaps you could stitch the torus back in a different way to get a three-dimensional manifold that is simply connected, but not a three-sphere. You could actually write out a presentation of the fundamental group, and then try to show that everything cancelled out to give just the unit element.

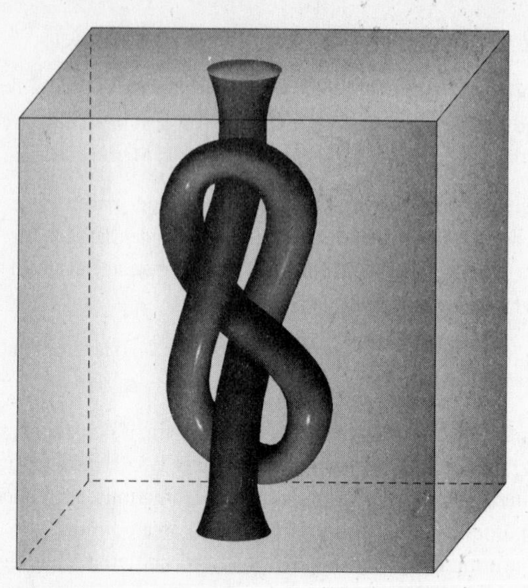

FIGURE 62. *Cube with a knotted hole*

The algebra was ferocious, but it was the sort of algebra that rewarded cleverness and that you could get lucky at. Once things canceled out, perhaps you could show somehow that the manifold was not homeomorphic to the three-sphere: maybe any function on it necessarily had more than two critical points. This is the sort of problem that makes you skip meals and not sleep.

214. The thesis was awarded upon the recommendation of Constantine Caratheodory for another proof of the invariance of homology groups of simplicial complexes, a result first proved by Alexander.

215. Dehn arrived in the United States in 1940, but had difficulty securing a permanent faculty position. He taught at the University of Idaho, Illinois Institute of Technology, and St. John's College, before winding up at Black Mountain College, an experimental institution in the mountains of North Carolina, and one especially famous for the arts. He died in 1952, a beloved faculty member, four years before the school closed its doors.

216. C. D. Papakyriakopoulos, "On Solid Tori," *Proceedings of the London Mathematical Society* 3, series 7 (1957): 281–99. (Let M be a 3-dimensional manifold with non-empty boundary ∂M and suppose that $f: S^1 \to \partial M$ is a closed loop (possibly with self intersections) which is homotopically zero in M but not ∂M, then there exists a simple (no self-intersections) loop $F: S^1 \to \partial M$ with the same property.) Papa proved the result under an orientablility assumption that was removed by John

Stallings, then a graduate student at Princeton, in J. Stallings, "On the Loop Theorem," *Annals of Mathematics* 72 (1960): 12–19. Dehn's lemma begins with the setup that the loop theorem guarantees that we can arrange. Namely, let *M* be a three-dimensional manifold with nonempty boundary ∂M and suppose that $f: S^1 \rightarrow \partial M$ is a simple closed loop (no self-intersections), which is homotopically zero in M. Then there is an embedding $F: D^2 \rightarrow M$ extending f. The proof appeared in C. D. Papakyriakopoulos, "On Dehn's Lemma and the Asphericity of Knots," *Annals of Mathematics* 66 (1957): 1–26. The phrase "asphericity of knots" refers to the sphere theorem that Papa proved in the same paper. Let M be a closed orientable three-dimensional manifold (so no boundary) and suppose that the second homotopy group $\pi_2 M$ is not trivial. Then there is an embedding $S^2 \rightarrow M$ which is homotopically nonzero. Papa's proof required some extra assumptions, which Henry Whitehead removed. (J. H. C. Whitehead, "On the sphere in 3-Manifolds," *Bulletin of the American Mathematical Society* 64 (1958): 161–66.) For a nice overview of this work, see C. D. Papakyriakopoulos, "Some Problems on 3-Dimensional Manifolds," *Bulletin of the American Mathematical Society* 64 (1958): 317–35.

217. R. H. Bing, "Necessary and Sufficient Conditions that a 3-Manifold be S^3," *Annals of Mathematics* 68 (1958) 17–37. K. Koseki, "Poincarésche Vermutung in Topologie," *Mathematics Journal of Okayama University* 8 (1958): 1–106.

218. C. D. Papakyriakopoulos, "A Reduction of the Poincaré Conjecture to Group Theoretic Conjectures," *Annals of Mathematics* 77 (1963): 250–305.

219. A. Doxiadis, *Uncle Petros and Goldbach's Conjecture* (New York, London: Bloomsbury, 1992, 2000).

220. R. H. Fox, "Construction of Simply Connected 3-Manifolds," in M. K. Fort, ed., *Topology of 3-Manifolds and Related Topics* (Englewood Cliffs, NJ: Prentice Hall, 1962), 213–16. R. H. Bing, "Some Aspects of the Topology of 3-Manifolds Related to the Poincaré Conjecture," in *Lectures on Modern Mathematics II*, ed. T. L. Saaty (New York: Wiley, 1964), 93–128.

221. J. R. Stallings, "How not to prove the Poincaré Conjecture," in *Topology Seminar Wisconsin, Annals of Mathematics Studies* 60 (1966): 83–88. Also available on his Web site (http://math.berkeley.edu/~stall).

222. If you have two manifolds M and N, the product of M and N, denoted $M \times N$, is the set of pairs (a,b) of elements, where a is in M and b is in N. $M \times N$ is a manifold. It is easy to see that the product of the two-dimensional sphere and the circle will be flat in one direction and curved in two others. Most of Thurston's extra geometries lived on products or spaces that that would be products if you could remove a surface from the manifold.

223. One shows that any simply connected three-dimensional manifold with spherical geometry is homeomorphic to the three-sphere. This is not difficult (see, for example, the books of Thurston or Weeks).

224. W. P. Thurston, *Three-Dimensional Geometry and Topology*, vol 1, edited by S. Levy, (Princeton: Princeton University Press, 1997). The set of Princeton lecture notes (W. P. Thurston, "The Geometry and Topology of Three-Manifolds") are on the Web, at the Web site of the Mathematical Sciences Research Institute in Berkeley: www.msri.org/publications/books/gt3m/. Both the books and the notes are wonderful.

225. W. P. Thurston, "Mathematical Education," *Notices of the American Mathematical Society* 37 (1990): 844–50.

226. See A. Jaffe and F. Quinn, "Theoretical Mathematics: Towards a Cultural Synthesis of Mathematics and Theoretical Physics," *Bulletin of the American Mathematical Society* 29 (1993): 1–13, and W. P. Thurston, "On Proof and Progress in Mathematics," *Bulletin of the American Mathematical Society* 30 (1994): 161–77.

227. Since it takes six different numbers to encode a mathematical object that assigns a number to every planar direction, the Ricci flow is given by $\partial g_{ij}/\partial t = -2R_{ij}$ is actually six equations, more in higher dimensions.

228. If one abandons the analogy with the heat equation, it is natural to ask about including the scalar curvature times the metric in the evolution equation. Einstein struggled with this question in formulating general relativity where again the only reasonable objects that can appear are the Ricci tensor and the scalar curvature times the metric. He first added in the latter to get a static universe, then took it out (calling its inclusion his biggest mistake) when it was discovered that the universe was expanding. Recent evidence suggests that it should be in.

229. R. S. Hamilton, "Three-Manifolds with Positive Ricci Curvature," *Journal of Differential Geometry* 17 (1982): 255–306.

230. D. M. DeTurck, "Deforming Metrics in the Direction of Their Ricci Tensors," *Journal of Differential Geometry* 18 (1983): 157–62. For a nice collection of papers on the Ricci flow, see H.-D. Cao, B. Chow, S.-C. Chu, and S.-T. Yau, eds., *Collected Papers on the Ricci Flow* (Somerville: International Press, 2003). An improved version of DeTurck's paper appears in this volume.

231. The critical case appeared in M. Gage and R. S. Hamilton, "The Heat Equation Shrinking Convex Plane Curves," *Journal of Differential Geometry* 23 (1986): 69–96. In M. Grayson, "The Heat Equation Shrinks Embedded Curves to Round Points,"

Journal of Differential Geometry 26 (1987): 285–314, it is shown that the flow in question makes any (embedded) closed curve convex, at which point the Gage-Hamilton result takes over.

232. This result is established for all cases except two spheres in which the curvature is zero at some point in R. S. Hamilton, "The Ricci Flow on Surfaces," *Contemporary Mathematics* 71 (1988): 237–61. The excluded case was established by B. Chow, "The Ricci Flow on the 2-Sphere," *Journal of Differential Geometry* 33 (1991): 325–34.

233. J. Birman. "Poincaré's Conjecture and the Homotopy Group of a Closed, Orientable 2-Manifold," *Journal of the Australian Mathematical Society* 17 (1974): 214–21.

234. The circumstances surrounding the choice of problems were sketched in the paper quoted in chapter 1 by the founding director, Arthur Jaffe, of the Clay Institute. The paper is entitled "The Millennium Grand Challenge in Mathematics," in *Notices of the American Mathematical Society,* 53, no. 6 (2006): 652–660.

235. From the video *The CMI Millennium Meeting Collection: Lectures by M. Atiyah, T. Gowers and J. Tate* (directed by F. Tisseyre), New York: Springer, 2002.

236. Königsberg was also the birthplace of Immanuel Kant. It is now called Kaliningrad, and was repopulated with Russian citizens following the Second World War. It was closed to foreign visitors in the 1950s. Following the breakup of the Soviet Union, and the admission of Poland and Lithuania into the European Union in 2004, it has been completely surrounded by the EU. There is now discussion of renaming the city.

237. Interest in the problems was even higher than anticipated: the large number of hits on Clay Institute's Web site brought down their Web server, and it looked as if the American Mathematical Society's higher-bandwidth server that had hosted a mirror site would also go down.

238. In soliciting problems, there were two that every mathematician listed: the Poincaré conjecture and the Riemann hypothesis. (See Jaffe.)

239. G. Perelman, "The Entropy Formula for the Ricci Flow and Its Geometric Applications," math.DG/0211159 (11 November 2002), "Ricci Flow with Surgery on Three-Manifolds," math.DG/0303109 (10 March 2003), "Finite Extinction Time for the Solutions to the Ricci Flow on Certain Three-Manifolds," math.DG/0307245 (17 July 2003).

240. G. Perelman, "The Entropy Formula," 2.

241. G. Perelman, "The Entropy Formula," 3.

242. Ibid.

243. Ibid.

244. G. Perelman, "The Entropy Formula," p. 4.

245. Leningrad Secondary School #239.

246. The other two were Bruno Haible, of Germany, and Lê Tu Quôc Thang, of Vietnam.

247. One requires that the manifold be noncompact and complete. The latter means that any boundaries (or missing points) are infinitely distant.

248. G. Perelman, "Proof of the Soul Conjecture of Cheeger and Gromoll," *Journal of Differential Geometry*, 40 (1994): 299–305.

249. The Steklov Institute is the mathematics division of the Russian Academy of Sciences, and was moved to Moscow in 1940. The Saint Petersburg Department occupies the original quarters, and there is a friendly, and sometimes sharp, rivalry between the Moscow and Saint Petersburg branches.

250. The incident is detailed early in her friend Alexandr Solzhenitsyn's book *The Gulag Archipelago* (New York: Harper and Row, 1973). Ladyzhenskaya was also a close friend of poet Anna Akhmatova. For a moving personal (and mathematical) appreciation of Ladyzhenskaya with many photographs, see S. Friedlander, P. Lax, C. Morawetz, L. Nirenberg, G. Seregin, N. Ural'tseva, and M. Vishik, "Olga Alexandrovna Ladyzhenskaya (1922–2004)," *Notices of the American Mathematical Society* 51, no. 11 (2004): 1320–1331. Ladyzhenskaya made decisive contributions to our understanding of the Navier-Stokes equations, the equations that govern fluid flow and that are, therefore, critical to weather prediction (our atmosphere is a fluid). These equations are the subject of yet another of the Clay millennium problems.

251. The exchange was forwarded to Don Davis at Lehigh, with instructions to post to the algebraic topology discussion group that Davis moderates.

252. Actually, he proves another conjecture of Thurston's, the *elliptization conjecture*, which states that every three-manifold with finite fundamental group has a metric with constant positive curvature. This is less general than the geometrization conjecture, but more general than the Poincaré conjecture, which follows immediately.

253. See www.math.lsa.umich.edu/research/ricciflow/perelman.html.

254. These have been submitted for publication and have been posted on www.arXiv.org. See B. Kleiner, J. Lott, "Notes on Perelman's Papers," math.DG/0605667, May 25, 2006 (192 pages).

255. At the time, Tian had an appointment at MIT.

256. The book has been submitted for publication and has been posted on www.arXiv.org. See J. W. Morgan, G. Tian, "Ricci Flow and the Poincaré Conjecture," math.DG/0607607, July 25, 2006 (473 pages).

257. G. Besson, "Une nouvelle approche de l'étude de la topologie des variétés de dimension 3 d'après R. Hamilton et G. Perel'man," *Séminaire Bourbaki*, 57ème année, 2004–05, no. 947, Juin 2005.

258. Both are students of Shing-Tung Yau, the Harvard mathematician whom we mentioned earlier, who received a Fields Medal in 1983 for his use of partial differential equations to establish the existence of metrics on various manifolds. Yau is one of the most influential mathematicians of current times and a master of the art of extracting geometric conclusions from partial differential equations. Xi-Ping Zhu is from Zhongshan University in Guangzhou, mainland China, and Cao is from Lehigh University. Yau and Cao continue to organize workshops on the Ricci flow (see, for example, www.math.harvard.edu/ricci/ricci.pdf).

259. H.-D. Cao, X.-P. Zhu, "A Complete Proof of the Poincaré and Geometrization Conjectures—Application of the Hamilton-Perelman Theory of the Ricci Flow," *Asian Journal of Mathematics* 10 no. 2 (June 2006): 165–492.

260. Brian Conrey, director of the American Institute of Mathematics, and David Eisenbud, director of the Mathematical Sciences Research Institute at Berkeley, both ran workshops on Perelman's work. In Europe, Jean-Pierre Bourguignon (director of the Institut des Hautes Études Scientifiques in France) and Gerhard Huisken (of the Albert-Einstein-Institut für Gravitationsphysik of the Max-Planck Society in Germany) convened study groups.

261. M. T. Anderson, "Geometrization of 3-Manifolds via the Ricci Flow," *Notices of the American Mathematical Society*, 51, no. 2 (2004): 184–193. L. Bessières, "Conjecture de Poincaré: la preuve de R. Hamilton et G. Perelman," *Gazette des Mathematiciens* 106 (2005): 7–35. J. W. Morgan, "Recent Progress on the Poincaré Conjecture and the Classification of 3-Manifolds," *Bulletin of the American Mathematical Society* 42 (2005): 57–78.

262. This conclusion emerges by combining observations from three different sources: the cosmic microwave background, type 1a supernova observations, and mass estimates.

263. The Abdus Salam International Center for Theoretical Physics in Trieste organizes a number of high-level meetings in Trieste each year devoted to specialized topics in theoretical physics and mathematics. In June 2005, the participants at the ICTP Summer School and Conference on Geometry and Topology of 3-Manifolds affirmed the proof. (See "Shapes, Spaces, and Spheres," *News from ICTP*, Summer 2005.) Needless to say, a vote of this type is very different from careful refereeing.

264. See the Xinhua news report from June 4, 2006 (news.xinhuanet.com/english/2006-06/04/content_4644754.htm).

265. The very first official press release for the congress in April, fully twenty weeks before the opening, announced that the Poincaré conjecture would be the theme of the Congress. In the same issue, the Spanish president of the executive committee, Manuel De León, speculated in an interview that Perelman's solution of the conjecture would be officially accepted during the congress. Subsequent press releases would return to the same theme.

266. Sylvia Nasar and David Gruber, "Manifold Destiny: A Legendary Problem and the Battle over Who Solved It," *The New Yorker* (August 28, 2006), 44–58.

267. Ball would later tell that press that he would leave to others the interpretation of the remarks, and that they were free to make their own interpretation.

268. The others went to Russian-born Andrei Okounkov of Princeton, Australian Terence Tao of UCLA, and German-born French mathematician Wendelin Werner of Paris-Sud and École Normale Supérieure.

269. "If I have seen further, it is by standing upon the shoulders of giants." Attributed to Isaac Newton (see, for example, *The Columbia World of Quotations*, 1996).

270. There was a little carping about the refereeing of the Cao-Zhu paper. No one doubts the correctness of their results. However, the *New Yorker* article had raised concerns about who had actually seen it.

Glossary of Terms

ATLAS: A collection of maps that covers the earth or the universe or a manifold.

BETTI NUMBERS: An integer counting that number of inequivalent submanifolds of a given dimension in a manifold that do not bound a submanifold of one dimension higher.

BOUNDARY: The edge of a manifold. If a manifold M has a boundary, then the boundary is a manifold of dimension one less than the dimension of M.

CLOSED PATH: A path (that is, a curve) on a manifold that begins and ends at the same point.

COMPACT: A manifold is compact if it has an atlas with a finite number of maps.

CONNECTED SUM: The manifold that results from cutting a solid ball out of each of two manifolds and identifying the points on the two spheres that bound the complements.

COMPLEX NUMBERS: Numbers obtained by augmenting the real numbers with the square roots of negative numbers.

COROLLARY: A proposition that follows easily from a theorem or another proposition.

CURVATURE: A mathematical object that measures the deviation of the sum of angles of triangles from 180 degrees. In a two-dimensional manifold, the curvature at each point is a number.

DIFFERENTIAL EQUATION: An equation in which one specifies rates of change of some unknown mathematical object and seeks the object as a solution.

DIMENSION: The number of independent degrees of freedom in a set. Equivalently, the minimum number of real numbers (that is, the number of coordinates) that it takes to specify a position near a given point in a set.

EUCLIDEAN SPACE: For each positive integer *n*, Euclidean space of dimension *n* is *n*-space with distance defined by the Pythagorean theorem.

FIFTH-POSTULATE: The fifth, and most complex, of the five postulates in book 1 of Euclid's *Elements*: If a straight line falling on two straight lines makes the interior angles on the same side less than two right angles (that is, whose sum is less than 180 degrees), the two straight lines, if extended indefinitely, meet on that side.

FLAT: A space is flat if the sum of the angles of every triangle in the space is 180 degrees.

GEOMETRIZATION CONJECTURE: The conjecture that every three-dimensional manifold can be cut up along spheres and tori into pieces that have one of the eight geometries.

GEODESIC: A curve that traces the shortest distance between any two of its points.

GEOMETRIC PROPERTY: A property that depends for definition on a distance or symmetry (for example: straight, angle measure, circle).

GEOMETRY: The structure that results from having a distance defined on a manifold.

GREAT CIRCLE: A circle on the (two-dimensional, round) sphere cut out by intersecting the sphere with a plane through its center.

GROUP (OF TRANSFORMATIONS): A set of transformations with the properties that (1) the result performing one transformation in the set, then another is a transformation that again is in the set and (2) the transformation that undoes a given transformation in the set is in the set. (That is, a group of a transformations is a set that is closed under the operations of taking products and forming inverses.)

HOMEOMORPHISM: A one-to-one correspondence between two manifolds in which nearby points correspond to nearby points.

LEMMA: A mathematical result whose main purpose and interest is as a stepping stone to proving another result.

LOOPS: A path that begins and ends at the same point. More technically, a continuous map of an interval into a manifold in which both endpoints get mapped to the same point.

MANIFOLD: A mathematical set that looks like Euclidean space at each point. (More formally, regions sufficiently near any point are homeomorphic to *n*-space.)

METRIC: A rule for specifying the distance between any two points of a set. In a manifold, a metric can be given by specifying a rule for measuring speed along curves.

N-SPACE: The set of all ordered *n*-tuples of real numbers.

NEGATIVE CURVATURE: A region in a manifold has negative curvature if the sums of all angles of all triangles in that region are less than 180 degrees.

POINCARÉ CONJECTURE: The conjecture, now proved, that every simply connected, compact three-dimensional manifold without boundary is homeomorphic to the three-dimensional sphere.

POSITIVE CURVATURE: A region in a manifold has positive curvature if the sums of angles of all triangles in that region are greater than 180 degrees.

PYTHAGOREAN THEOREM: The statement that the square of the length of the hypotenuse of a right-angle triangle is equal to the sum of the squares of the two sides meeting at the right angle. (The hypotenuse is the side across from the right angle.)

RICCI FLOW: The process governed by the Ricci flow equation that specifies that the metric on a manifold changes by having curvature flow from more highly curved to less highly curved areas. The Ricci flow equation is actually a system of equations denoted symbolically by $\partial g_{ij}/\partial t = -2 R_{ij}$.

RIEMANN CURVATURE TENSOR: A mathematical object that assigns a value to every planar direction through the point (which reflects how much angle sums of tiny geodesic triangles in that direction tend to deviate from 180 degrees).

ROUND SPHERE: The word *round* refers to a sphere which has the same curvature at every point and is used when one wants to make a distinction between such spheres and topologically equivalent spheres that may have bumps. The surface of our earth is not a round sphere because it is flattened at the poles. A round sphere is isometric to the set of points at some fixed distance from a point in Euclidean space.

SIMPLY CONNECTED: A manifold is simply connected if any loop can be shrunk to a point. This is equivalent to the statement that the fundamental group consists of a single element (necessarily, the identity).

SPHERE: Without additional qualification, refers to a two-dimensional sphere which is any object homeomorphic to the set of points in three-space at a fixed distance from a given point. The surface of a ball is a sphere. There are, however, spheres of every dimension. The simplest definition of a sphere of a given dimension is as the set of points of fixed distance from a single point in the Euclidean space of one dimension higher. So, for example, the set of points at distance one from a fixed point in six-dimensional Euclidean space is a five-dimensional sphere.

PARTIAL DIFFERENTIAL EQUATION: A type of differential equation in which one specifies rates of change at different points in different directions. The solutions of partial differential equations are objects that have the desired rates of change at all points in all directions. Many equations of mathematical physics are partial differential equations.

POSTULATE: An assertion accepted without proof. Synonomous with axiom.

PROOF: A complete rigorous argument in which each assertion is an axiom or previously proved proposition, or else follows from such by formal rules of logic. It begins with axioms and known propositions and ends with the statement that is to be proved.

PROPOSITION: A statement that is derived from postulates and previously proved propositions using mathematical reasoning.

RICCI FLOW: The evolution of the metric on a manifold in accordance with the Ricci flow equation. The latter is actually a set of equations that specify that curvature changes by decreasing in directions along which the curvature is greater and increasing in directions along which the curvature is less. Symbolically, $\partial g_{ij}/\partial t = -2 R_{ij}$.

SURFACE: A two-dimensional manifold.

TENSOR: A mathematical object that assigns real numbers to a prescribed number of vectors (that is, velocities) at each point of a manifold.

THEOREM: An especially important proposition.

THREE-DIMENSIONAL MANIFOLD (OR THREE-MANIFOLD): An idealized mathematical shape that models shapes that three-dimensional spaces, like our universe, might have. The region around every point can be mapped onto the inside of a solid aquarium. Put differently, the region near every point looks like three-space.

THREE-SPHERE: A manifold constructed by taking two solid balls and matching up the points on the (spherical) boundary of each. The set of points at a fixed distance from a point in four-dimensional space is a three-sphere.

THREE-TORUS: The manifold constructed by connecting the opposite faces of a solid rectangular box.

TOPOLOGICAL PROPERTY: A property that is invariant under continuous homeomorphisms. Examples are connectivity, simple connectedness, dimension.

TOPOLOGICALLY EQUIVALENT: Two manifolds are topologically equivalent if they are homeomorphic.

TOPOLOGY: The study of shapes.

TORUS: The surface of a doughnut.

TWO-DIMENSIONAL MANIFOLD (OR TWO-MANIFOLD): An idealized mathematical shape that models the surfaces of possible worlds. The region around every point can be mapped onto a sheet of paper (that is, the inside of a rectangle in the plane).

TWO-SPACE: The plane envisaged by Euclid that goes on forever in two independent directions. As a set, it is just the set of pairs of real numbers.

Glossary of Names

Pavel Aleksandrov (1896–1982): Russian topologist who made fundamental contributions to homology theory. Wrote a now classic book with Heinz Hopf.

James Alexander (1888–1971): Princeton topologist who significantly advanced three-dimensional topology and the study of the Poincaré conjecture.

Al-Hajjaj (c. 786–833): Arab translator of Theon's Euclid.

Al-Idrisi (c. 1100–1165): Great geographer, descendant of the Prophet, in the court of Roger II.

Al-Mansur (712–75): Second Abbasid Caliph: reigned in Baghdad from 754–75 and obtained Theon's Euclid from the Byzantine emperor.

Friedrich Althoff (1839–1908): Prussian Minister of Education.

Archytas (428–350 bce): Pythagorean mathematician.

Aristotle (384–322 bce): Great philosopher, tutor to Alexander the Great.

Martin Bartels (1769–1833): Assistant to Gauss's grade-school teacher, and subsequently a friend of Gauss. He later became a mathematics professor at the University of Kazan, where he interested Lobachevsky in mathematics.

Eugenio Beltrami (1835–1900): Italian geometer who was the first to realize that the behavior of lines in Lobachevskian geometry was the same as the behavior of geodesics on surfaces of constant negative curvature.

George D. Birkhoff (1884–1944): One of the first American-born mathematicians to achieve international fame. Built up the Department of Mathematics at Harvard. Student with Veblen of E. H. Moore.

FARKAS WOLFGANG BOLYAI (1775–1856): Mathematician friend of Gauss. Father of János Bolyai.

JÁNOS BOLYAI (1802–60): Hungarian mathematician who independently developed non-Euclidean geometry.

WILLIAM KINGDON CLIFFORD (1845–79): Gifted English geometer who speculated that Riemann's work could be used to develop a geometric theory of gravitation.

RICHARD COURANT (1882–1972): Applied mathematician. Successor to Felix Klein and head of the Mathematics Institute at Göttingen. He founded the great mathematics department at New York University (NYU).

AUGUST LEOPOLD CRELLE (1780–1855): Prussian civil engineer who loved mathematics, supported many younger mathematicians, and founded a famous journal.

DANTE ALIGHIERI (1261–1321): Poet, diplomat, and author of *Divine Comedy*, in which the universe is conceptualized as a three-sphere.

GASTON DARBOUX (1842–1917): Celebrated French geometer, head of Poincaré's Ph.D. thesis committee; delivered Poincaré's eulogy to the French Academy of Science.

RICHARD DEDEKIND (1831–1916): Accomplished algebraic number theorist, friend of Riemann, and successor to Gauss's chair at the University of Göttingen from 1855.

MAX DEHN (1878–1952): Topologist who made decisive contributions to three-manifold topology and knot theory. He wrote one of the first tracts on topology to discuss Poincaré's work.

JOHANN PETER GUSTAV LEJEUNE DIRICHLET (1805–59): Belgian-born number theorist, trained in Paris, who was one of the intellectual leaders in Berlin mathematics before moving to Göttingen. Married to composer Felix Mendelssohn's sister, he stressed conceptual thought over computation and was an enormous influence on Riemann.

FERDINAND GOTTHOLD MAX EISENSTEIN (1823–52): Brilliant number theorist at the University of Berlin.

ERATOSTHENES (275–195 BCE): hird head librarian of Alexandria library. Mathematician and geographer who most accurately estimated the Earth's circumference.

EUCLID (C. 325–C. 265 BCE): Greek geometer associated with Alexandria library. Author of the *Elements*.

RALPH FOX (1913–73): Princeton knot theorist, student of Lefschetz, advisor of John Milnor, John Stallings, and Barry Mazur. Brought Papakyriakopoulus to Princeton.

JOHANN CARL FRIEDRICH GAUSS (1777–1855): One of the greatest mathematicians of all time. The Göttingen professor who discovered non-Euclidean geometry and advanced the study of differential geometry of surfaces.

GHERARD OF CREMONA (1114–87): Italian mathematician who worked in Toledo, Spain, and translated over eighty mathematical and scientific texts from Arabic into Latin. He first translated the Arabic translation of Theon's Euclid back into Latin.

RICHARD HAMILTON (1983–): American mathematician, student of Robert Gunning of Princeton. Introduced the Ricci flow equation and used it to study geometrization.

POUL HEEGAARD (1871–1948): Danish topologist whose thesis pointed out a mistake in Poincaré's first topological paper.

DAVID HILBERT (1862–1943): Leading mathematician at the turn of the last century, second only to Poincaré.

CARL GUSTAV JACOB JACOBI (1804–51): One of the most computationally gifted mathematicians of all time. Central to the Berlin school of mathematics.

JULIUS CAESAR (100–44 BCE): Roman leader who burned Alexandria.

ABRAHAM KÄSTNER (1719–1800): Mathematics professor at the University of Göttingen who was interested in the fifth postulate and taught a course on geometry that Gauss took as a student.

FELIX KLEIN (1849–1925): Charismatic mathematician and administrator who built the mathematics department at the University of Göttingen.

ANDREI KOLMOGOROV (1903–1987): One of the most famous mathematicians of all time. Contributed to all areas of mathematics. A Russian Poincaré. Discovered cohomology theory.

JOHANN HEINRICH LAMBERT (1728–77): French mathematician who worked at the Prussian Academy of Sciences in Berlin. He wrote an influential book, published posthumously, carefully exploring the consequences of assuming that the fifth postulate is false.

HYPATIA (370–415 CE): Famed Neo-platonist mathematician, daughter of Theon.

IMMANUEL KANT (1724–1804): Last great Enlightenment philosopher.

SOLOMON LEFSCHETZ (1884–1972): Colorful, Russian-born mathematician who became famous applying Poincaré's work to algebraic geometry while working at the University of Kansas. He was a major contributor to the excellence of the Princeton Mathematics Department, the journal *Annals of Mathematics,* and American mathematics.

ADRIEN-MARIE LEGENDRE (1752–1833): Great French mathematician whose modernization of Euclid in 1794 became the leading text on geometry for almost a century.

TULLIO LEVI-CIVITA (1873–1941): Italian mathematician, student of Ricci, who developed Riemannian geometry and tensor calculus into a form suitable for applications.

NIKOLAI IVANOVICH LOBACHEVSKY (1792–1856): Mathematics professor and rector of the University of Kazan. One of the founders of non-Euclidean (hyperbolic) geometry.

NIKOLAI NIKOLAEVICH LUZIN (1883–1950): One of the founders of the twentieth-century Russian school of mathematics. Supervisor of Aleksandrov and Kolmogorov.

MARSTON MORSE (1892–1977): Student of Birkhoff, member of Institute for Advanced Study, and developer of Morse theory.

JOHN MILNOR (1931–): American mathematician whose discovery of different differentiable structures on the seven-dimensional sphere triggered a new era in differential topology.

CHRISTOS PAPAKYRIAKOPOULOS (1914–1976): Greek mathematician who devoted his life to the Poincaré conjecture.

PHILOLAUS (470–385 BCE): Pythagorean mathematician whose book contributed to the wide knowledge of Pythagorean teachings in the ancient world.

MAXIMOS PLANUDES (C. 1260–1330): Byzantine monk, rediscovered Ptolemy's *Geography*.

PLATO (427–347 BCE): Great Greek philosopher, founder of the Academy.

PROCLUS (411–485 CE): Neoplatonist philosopher who became head of Plato's Academy in Athens and who wrote a commentary on Euclid's *Elements*.

PTOLEMY I SOTER (367–282 BCE): One of Alexander the Great's ablest generals, and founder of the great Museum and Library at Alexandria. He was appointed satrap of Egypt following Alexander's death in 323 BCE and declared himself King Ptolemy I in 305 BCE.

CLAUDIUS PTOLEMY (85–165 CE): Greatest geographer and astronomer of Alexandria. He wrote a book, now lost, attempting to prove the fifth postulate.

PYTHAGORAS OF SAMOS (569–475 BCE): Great Greek geometer and philosopher. Founder of the influential Pythagorean school.

GREGORIO RICCI-CURBASTRO (1853–1925): Italian mathematician who, with his student Levi-Civita, developed tensor calculus to express Riemannian geometry. The Ricci tensor is named after him.

GEORGE FRIEDRICH BERNHARD RIEMANN (1826–66): Gloriously creative German mathematician. One of the greatest mathematicians of all time, he made fundamental contributions to analysis, number theory, and geometry. Einstein's general relativity depends crucially on his ideas.

ROGER II (1093–1154): Tolerant Norman king of Sicily with famed multiethnic court.

GEROLAMA SACCHERI (1667–1733): Jesuit mathematician who made substantial progress on understanding the fifth postulate.

STEPHEN SMALE (1930–): Berkeley professor (now at City University of Hong Kong) who won the Fields Medal for his proof of the analog of the Poincaré conjecture in all dimensions greater than or equal to five.

JAKOB STEINER (1796–1863): Mathematics professor at the University of Berlin, especially known for his work in projective geometry,

THEON (335–C. 405 CE): Late Alexandrian mathematician, whose edition of Euclid with his daughter was translated into Arabic.

RENÉ THOM (1923–2002): French mathematician who won Fields Medal in 1958 for his work in cobordism theory. He founded catastrophe theory.

WILLIAM THURSTON (1946–): American geometer and Field Medalist. He identified the eight geometries in three dimensions and framed the geometrization conjecture.

HEINRICH TIETZE (1880–1964): Austrian topologist who was a professor at the University of Munich from 1925 on and who clarified Poincaré's work.

OSWALD VEBLEN (1880–1960): American mathematician influenced by work of Poincaré. He built the strong topological tradition at Princeton.

ALEXANDER VON HUMBOLDT (1769–1859): World-famous scientist and explorer who did much to build Berlin mathematics. Brother of Wilhelm.

WILHELM VON HUMBOLDT (1767–1835): Minister of education in Prussia from 1809, founder of the University of Berlin in 1810. He oversaw the development of an education system aimed at providing a rigorous education to all social classes.

WILHELM EDUARD WEBER (1804–91): Accomplished physicist, and collaborator of and son-in-law to Gauss.

J. H. C. WHITEHEAD (1904–60): British mathematician who trained at Princeton and began the strong tradition of topology at Oxford. He announced a mistaken proof of the Poincaré conjecture.

SHING-TUNG YAU (1949–): Fields Medalist in 1983 for his work using partial differential equations to find canonical metrics on spaces. Early proponent of Ricci flow methods.

Timeline

The narrative develops the following three interlocking themes.

1. The development of geometry and topology underlying, and pertaining to, the Poincaré Conjecture:

   Babylonians → Pythagoreans → Alexandrians (esp. Euclid) → Arab translations → Gauss/Lobachevsky/Bolyai → Riemann → Poincaré → Göttingen/Moscow/Princeton → Higher dimensions → Thurston → Partial differential equations → Hamilton/Perelman

2. The evolving question of the shape of the universe:

   Egyptians → Greeks → Columbus and explorers → Riemann → Einstein → Weeks

3. The evolving social context of mathematics:

   Babylonians → Greek academies → Arab courts → Early universities → Scientific societies → German research university → National universities → U.S. and Soviet universities → Mathematical organizations and the Internet

Each of these themes unfolds against a larger background of historical events and people. Here is a *very* incomplete timeline to help orient the reader. The first column lists dates (mentioned in the text) from Babylonian times to present. A number of events (mostly mathematical discoveries) pertaining to themes 1 and 2 have been listed in the second column. Historical events of particular importance to the narrative are listed in the third column. Events pertaining to theme 3, such as conferences, or the founding of institutions, are listed in the fourth column.

DATE	MATHEMATICS	POLITICAL EVENTS	INSTITUTIONS
1700 BCE	Triangle geometry	Hammurabi and first Babylonian dynasty	
547 BCE	Number theory origins and early proofs	Ionian philosopher Thales dies	
530 BCE		Pythagoras and followers move to Croton	Pythagorean semicircle
387 BCE			Plato's Academy founded
323 BCE		Alexander dies. His general, Ptolemy, takes power in Egypt.	Alexandria library founded
300 BCE	Euclid's *Elements*		
240 BCE	Erastosthenes estimates Earth's circumference		
47 BCE		Caesar burns Alexandria harbor	
150 CE	Ptolemy's *Geography*		
410		Rome sacked by Alaric	
415		Hypatia murdered	
529			Plato's Academy closed by Justinian
1088			University of Bologna recognized
1117			Oxford recognized
1144	Gherard's retranslation of Greek math and science books begins		

DATE	MATHEMATICS	POLITICAL EVENTS	INSTITUTIONS
1150			University of Paris starts
1187	Gherard dies in Toledo		
1209			Cambridge University recognized
1320	Dante's *Paradiso*: the universe as a three-sphere		
1492		Columbus's voyage	
1660			Royal Society (London) formed
1636			Harvard founded
1666			Academie des Sciences (Paris) formed
1724			Russian/St. Petersburg Academy of Science formed
1737			University of Göttingen founded
1746			Princeton founded
1789		French Revolution	
1794	Legendre's *Geometry*		
1799		Napoleon seizes power	
1810			Humboldt opens University of Berlin
1814		Congress of Vienna	
1824	Gauss and Bolyai discover non-Euclidean geometry		

(continued)

(continued)

DATE	MATHEMATICS	POLITICAL EVENTS	INSTITUTIONS
1829	Lobachevsky publishes in Russian in local journal		
1837		Victoria ascends throne; Göttingen Seven dismissed	Mount Holyoke, women's college, opens doors
1848		German Revolution; Klein born	
1854	Riemann's Habilitation lecture	Poincaré born	
1866	Riemann shows topology and analysis inextricably linked	Riemann dies	
1870		Franco-Prussian War	
1880	Poincaré realizes non-Euclidean geometry at heart of what he was studying		
1881	Klein–Poincaré letters		
1884			Abbott's *Flatland* appears
1886			Klein gets offer from Johns Hopkins; he moves to Göttingen.
1892			University of Chicago opens. E. H. Moore head of mathematics department
1895	Poincaré's first topological paper		Hilbert accepts job at Gottingen
1900	Hilbert's twenty-three problems		

DATE	MATHEMATICS	POLITICAL EVENTS	INSTITUTIONS
1904	Poincaré's fifth complement		
1905	Einstein's theory of Special Relativity		Veblen hired at Princeton
1908	Dehn submits and withdraws proof of Poincaré conjecture		
1912	Poincaré dies		Birkhoff hired at Harvard
1914–18	Einstein's theory of general relativity (1915)	First World War	Luzin hired at Moscow
1932		Hitler wins plurality in German elections	Institute for Advanced Study at Princeton opens
1934	Whitehead publishes proof of Poincaré. Seifert-Threlfall text appears.		
1935	Whitehead withdraws proof. Alexandroff-Hopf text appears. Cohomology announced.		International topology conference at Moscow
1939–45		Second World War	
1950			National Science Foundation in U.S. founded
1956	Milnor's exotic spheres. Nash solves Riemann embedding problem.		
1960	Smale proves Poincaré conjecture in dimensions five and more		

(continued)

(continued)

DATE	MATHEMATICS	POLITICAL EVENTS	INSTITUTIONS
1982	Freedman proves Poincaré in dimension four. Hamilton uses Ricci flow.	1982 International Congress of Mathematicians (ICM) postponed	Perelman gets medal for perfect score in International Mathematical Olympiad in Budapest. MSRI founded.
1983	Thurston's work on geometries and foliations and Yau's on geometric analysis recognized by Fields Medals		ICM in Warsaw
1986	Rourke, Rego announce proof of Poincaré		
1994	Perelman proves soul conjecture		
1995	Poénaru program		
1998	McMullen receives Fields Medal partly for work on hyperbolic three-manifolds		ICM in Berlin
2000	Millennium problems announced		
2001	Wilkinson Microwave Anisotropy Probe launched	9/11 terrorist attack	
2002	Perelman's first post to arXiv.org		ICM in Beijing
2003	Perelman's MIT and Stony Brook lectures		
2006	Perelman receives Fields Medal		ICM in Madrid

Bibliography

Abbott, E. A. *Flatland: A Romance of Many Dimensions.* Princeton: Princeton University Press, 1991.

Adams, C. *The Knot Book.* Providence: The American Mathematical Society, 2004.

Abdali, S. K. The Correct Qibla. http://www.patriot.net/users/abdali/ftp/qibla.pdf.

Alexander, J. W. "Note on Two 3-Dimensional Manifolds with the Same Group." *Transactions of the American Mathematical Society* 20 (1919): 339–42.

———. "Some Problems in Topology." *Verhandlungen des Internationalen Mathematiker Kongresses Zürich* (1932); Kraus reprint (1967): 249–57.

Anderson, M. T. "Geometrization of 3-Manifolds via the Ricci Flow." *Notices of the American Mathematical Society* 51, no. 2 (2004): 184–193.

Arnold, V. I. "On Teaching Mathematics." *Russian Mathematical Surveys* 53, no. 1 (1998): 229–36.

Atiyah, M. et al. Responses to "Theoretical Mathematics: Toward a Cultural Synthesis of Mathematics and Theoretical Physics" by A. Jaffe and F. Quinn, *Bulletin of the American Mathematical Society* 30 (1994): 178–207.

Barden, D. and C. Thomas, *An Introduction to Differential Manifolds.* London: Imperial College Press, 2003.

Beardon, A. *The Geometry of the Discrete Groups.* New York: Springer-Verlag, 1983.

Bell, E. T. *Men of Mathematics,* New York: Simon and Schuster, 1937.

Berger, M. *A Panoramic View of Riemannian Geometry.* Berlin: Springer-Verlag, 2003.

Bessières, L. "Conjecture de Poincaré: La Preuve de R. Hamilton et G. Perelman." *Gazette des Mathématiciens* 106 (2005): 7–35.

Bing, R H. "Necessary and Sufficient Conditions that a 3-Manifold Be S^3." *Annals of Mathematics* 68 (1958): 17–37.

———. "Some Aspects of the Topology of 3-Manifolds Related to the Poincaré Conjecture." In Saaty, T. L., ed., *Lectures on Modern Mathematics II*. New York: Wiley, 1964, 93–128.

Birkhoff, G. D. "Fifty Years of American Mathematics." *Science* 88, no. 2290 (1938): 461–67.

———. "Topology." *Science* 96, no. 2,504 (1942): 581–84.

Birman, J. "Poincaré's Conjecture and the Homotopy Group of a Closed, Orientable 2-Manifold." *Journal of the Australian Mathematical Society* 17 (1974): 214–21.

Bonola, R. *Non-Euclidean Geometry: A Critical and Historical Study of Its Development*. Translated by H. S. Carslaw. New York: Dover, 1955.

Bottazzini, U. *Henri Poincaré: Philosophe et mathematicien*. Paris: Pour la Science, 2000.

Brittain, V. *Testament of Youth: An Autobiographical Study of the Years 1900–1925*. New York: Penguin, 1989 (first published in London, 1933).

Browder, F. E. *The Mathematical Heritage of Henri Poincaré* (2 vols.). Providence: American Mathematical Society, 1983.

Bühler, W. K. *Gauss: A Biographical Study*. Berlin, New York: Springer-Verlag, 1981.

Callahan, J. J. "The Curvature of Space in a Finite Universe." *Scientific American* 235, August 1976, 90–100.

Cannon, J. W., W. J. Floyd, R. Kenyon, and W. R. Parry. "Hyperbolic Geometry," In *Flavors of Geometry*, edited by S. Levy. Cambridge: Cambridge University Press, 1997.

Cao, H.-D., B. Chow, S. C. Chu, and S. T. Yau, eds. *Collected Papers on the Ricci Flow*. Somerville: International Press, 2003.

Cao, H.-D. and X.-P. Zhu. "A Complete Proof of the Poincaré and Geometrization Conjectures—Application of the Hamilton-Perelman Theory of the Ricci Flow." *Asian Journal of Mathematics* 10 (2006): 165–492.

Chow, B. "The Ricci Flow on the 2-Sphere." *Journal of Differential Geometry* 33 (1991): 325–34.

Clifford, W. K. *Mathematical Papers*, edited by R. Tucker. London: Macmillan, 1882.

Collins, J. *Good to Great: Why Some Companies Make the Leap . . . and Others Don't*. New York: HarperCollins, 2001.

Dante Alighieri. *The Divine Comedy*. (c. 1318) Translated by A. Mandelbaum. New York: Knopf, 1995.

Darboux, G. "Éloge Historique d'Henri Poincaré." In Poincaré, Henri. *Œuvres*, vol. 2. Paris: Gauthier-Villars, 1916.

Dehn, M., and P. Heegard. "Analysis Situs," in *Enzyklopädie der Mathematischen Wissenschaften III*, AB 3. Leipzig: Teubner, (1907): 153–220.

Dehn, M. *Papers on Group Theory and Topology*, translated by and with an introduction by J. Stillwell. New York: Springer-Verlag, 1987.

Derbyshire, J. *Prime Obsession: Bernhard Riemann and the Greatest Unsolved Problem in Mathematics.* Washington: Joseph Henry Press, 2001.

DeTurck, D. M. "Deforming Metrics in the Direction of Their Ricci Tensors." *Journal of Differential Geometry* 18 (1983): 157–62.

Doxiadis, A. *Uncle Petros and Goldbach's Conjecture.* New York, London: Bloomsbury, 1992, 2000.

du Sautoy, M. *The Music of the Primes: Searching to Solve the Greatest Mystery in Mathematics.* New York: HarperCollins, 2003.

Dupont, J. L., and C.-H. Sah. "Scissors Congruences." *Journal of Pure and Applied Algebra* 25 (1982): 159–95.

Duren, P. *A Century of Mathematics in America.* 3 vols. Providence: American Mathematical Society, 1989.

Durfee, A. H. "Singularities," in *History of Topology*, edited by I. M. James. Amsterdam: Elsevier, 1999, 417–34.

Ewald, W. B. *From Kant to Hilbert: A Source Book in the Foundations of Mathematics.* New York: Oxford University Press, 1996, vol. 1.

Epple, M. "Geometric Aspects in the Development of Knot Theory," in *History of Topology*, edited by I. M. James. Amsterdam: Elsevier, 1999, 301–58.

Feffer, L. B. "Oswald Veblen and the Capitalization of American Mathematics: Raising Money for Research, 1923–1928," *Isis* 89 (1998): 474–97.

Fox, R. H. "Construction of Simply Connected 3-Manifolds," in *Topology of 3-Manifolds and Related Topics*, edited by M. K. Fort. Englewood Cliffs, NJ: Prentice Hall, 1962, 213–16.

Freedman, M. "The topology of Four Manifolds." *Journal of Differential Geometry* 17 (1982): 357–454.

Friedlander, S., P. Lax, C. Morawetz, L. Nirenberg, G. Seregin, N. Ural'tseva, and M. Vishik. "Olga Alexandrovna Ladyzhenskaya (1922–2004)." *Notices of the American Mathematical Society* 51, no. 11 (2004): 1320–31.

Gabai, D. "Valentin Poénaru's Program for the Poincaré Conjecture," in *Geometry, Topology and Physics for Raoul Bott*, edited by S.-T. Yau. Cambridge: International Press, 1994, 139–66.

Gage, M., and R. S. Hamilton. "The Heat Equation Shrinking Convex Plane Curves." *Journal of Differential Geometry* 23 (1986): 69–96.

Galison, P. *Einstein's Clocks, Poincaré's Maps: Empires of Time.* New York: W. W. Norton, 2003.

Gallot, S., D. Hulin, and J. Lafontaine. *Riemannian Geometry,* 3rd ed. Berlin: Springer, 2004.

Gauss, C. F. *Werke.* Göttingen: K. Gesellschaft der Wissenschaften zu Göttingen, 1863–1933.

Gillespie, C. C. ed. *Dictionary of Scientific Biography.* New York: Scribner, 1971.

Gillman, D., and D. Rolfsen. "The Zeeman conjecture for Standard Spines Is Equivalent to the Poincaré Conjecture." *Topology* 22 (1983): 315–23.

Gleick, J. *Chaos: Making a New Science.* New York: Penguin, 1988.

Goldberg, L., and A. V. Phillips, eds. *Topological Methods in Modern Mathematics.* Houston: Publish or Perish Press, 1993.

Gordon, C. McA. "3-Dimensional Topology up to 1960," in *History of Topology*, edited by I. M. James. Amsterdam: Elsevier, 1999, 449–90.

Gorman, P. *Pythagoras: A Life.* London, Boston: Routledge and K. Paul, 1979.

Gray, J. *Ideas of Space: Euclidean, Non-Euclidean, and Relativistic.* Oxford, New York: Oxford University Press, 1979.

Gray, J. J., and S. A. Walter, eds. *Henri Poincaré, Three Supplementary Essays on the Discovery of Fuchsian Functions.* Berlin: Akademie Verlag GmbH, and Paris: Albert Blanchard, 1997.

Grayson, M. "The Heat Equation Shrinks Embedded Curves to Round Points." *Journal of Differential Geometry* 26 (1987): 285–314.

Greffe, J. L., G. Heinzmann, and K. Lorenz, eds. *Henri Poincaré, Science and Philosophy.* Berlin: Akademie-Verlag GmbH, and Paris: Albert Blanchard, 1996, 241–50.

Hadamard, J. "L'Oeuvre Mathématique de Poincaré." *Acta Mathematica* 38 (1921): 203–87.

———. *Non-Euclidean Geometry in the Theory of Automorphic Functions* (1951), translated by A. Shenitzer, edited by J. J. Gray and A. Shenitzer. Providence: American Mathematical Society, 1999.

Haken, W. "Some Results on Surfaces in 3-Manifolds," in *Studies in Modern Topology*, edited by P. Hilton. MAA Studies. Washington: Mathematical Association of America.

Halsted, G. B. "Biography, Bolyai Farkas [Wolfgang Bolyai]." *American Mathematical Monthly* 3 (1896): 1–5.

Hamilton, R. S. "Three-Manifolds with Positive Ricci Curvature." *Journal of Differential Geometry* 17 (1982): 255–306.

———. "The Ricci Flow on Surfaces." *Contemporary Mathematics* 71 (1988): 237–61.

———. "The Formation of Singularities in the Ricci Flow." *Surveys in Differential Geometry* 2 (1995): 7–136.

Haskins, C. H. *The Rise of Universities.* New York: Gordon Press, 1957.

Heath, T. L. *A History of Greek Mathematics.* 2 vols. Oxford, 1921.

———. *The Thirteen Books of Euclid's Elements.* New York: Dover, 1956.

Heinzmann, G. "Éléments preparatoire à une biographie d'Henri Poincaré," preprint.

Hempel, J. *3-Manifolds.* Princeton: Princeton University Press, 1976.

Heyerdahl, T. *Early Man and the Ocean: A Search for the Beginnings of Navigation and Seaborne Civilizations.* Garden City, NY: Doubleday, 1979.

Hilbert, D. *Die Grundlagen der Geometrie.* Leipzig: Teubner, 1899.

Irving, W. *Life and Voyages of Columbus.* London: John Murray, 1830.

Jaffe, A. "The Millennium Grand Challenge in Mathematics." *Notices of the American Mathematical Society* 53 (2006): 652–60.

Jaffe A., and F. Quinn. "Theoretical mathematics: Towards a Cultural Synthesis of Mathematics and Theoretical Physics." *Bulletin of the American Mathematical Society* 29 (1993): 1–13.

———. Response to comments on "Theoretical mathematics." *Bulletin of the American Mathematical Society* 30 (1994): 208–11.

Jakbsche, W. "The Bing-Borsuk Conjecture Is Stronger than the Poincaré Conjecture." *Fundamenta Mathematicae* 106 (1980): 127–34.

James, I. M. ed. *History of Topology.* Amsterdam: Elsevier, 1999.

Johns, A. *The Nature of the Book: Print and Knowledge in the Making.* Chicago: University of Chicago Press, 1998.

Kagan, V. N. *Lobachevsky and His Contribution to Science.* Moscow: Foreign Languages Publishing House, 1957.

Kervaire, M. "A Manifold which Does Not Admit and Differentiable Structure." *Commentarii Mathematici Helvetici* 34 (1960): 257–70.

Klein, F. *Vorlesungen über die Entwicklung der Mathematik im 19 Jahrhundert.* Teil I, II. Berlin: Springer, 1926 (Teil I), 1927 (Teil II).

Kleiner, B. and J. Lott, eds. Web masters. "Notes and Commentaries on Perelman's Ricci Flow Papers," http://www.math.lsa.umich.edu/~lott/ricciflow/perelman.html.

———. "Notes on Perelman's Papers." arXiv:math.DG/0605667v1, 25 May 2006.

Kolmogorov, A. "The Moscow School of Topology." *Science* 97 no. 2,530 (1943): 579–80.

Koseki, K. "Poincarésche Vermutung in Topologie." *Mathematical Journal of Okayama University* 8 (1958): 1–106.

Kosinski, A. *Differential Manifolds.* New York: Academic Press, 1993.

Laugwitz, D. *Bernhard Riemann 1826–1866: Turning Points in the Conception of Mathematics.* Boston: Birkauser, 1999.

Lefschetz, S. "Reminiscences of a Mathematical Immigrant in the United States." *American Mathematical Monthly* 77 (1970): 344–50.

———. "James Waddell Alexander (1888–1971)." In *Yearbook of the American Philosophical Society (1973),* Philadelphia, 1974, 110–14.

Mawhin, J. "Henri Poincaré. A Life in the Service of Science." *Notices of the American Mathematical Society* 52 (2005): 1036–44.

McMullen, C. "Riemann Surfaces and the Geometrization of 3-Manifolds." *Bulletin of the American Mathematical Society* 27 (1992): 207–16.

Milnor, J. "On the Total Curvature of Knots." *Annals of Mathematics* 52 (1950): 248–57.

———. "On Manifolds Homeomorphic to the 7-Sphere." *Annals of Mathematics* 64 (1956): 399–405.

———. *Singular Points of Complex Hypersurfaces.* Princeton: Princeton University Press, 1968.

———. "Towards the Poincaré Conjecture and the Classification of Three-Manifolds." *Notices of the American Mathematical Society* 50, no. 10 (2003): 1226–33.

———. "The Poincaré Conjecture One Hundred Years Later." http://www.math.sunysb.edu/~jack.

Monastyrsky, M. *Modern Mathematics in the Light of the Fields Medals.* Wellesley: AK Peters, 1998.

Morgan, J. W. "Recent Progress on the Poincaré Conjecture and the Classification of 3-Manifolds." *Bulletin of the American Mathematical Society* 42 (2005): 57–78.

Nakayama, S. *Academic and Scientific Traditions in China, Japan and the West.* Translated by J. Dusenbury. Tokyo: University of Tokyo Press, 1984.

Nasar, S. *A Beautiful Mind.* New York: Simon and Schuster, 1998.

Nash, J. "The Imbedding Problem for Riemannian Manifolds." *Annals of Mathematics* 63 (1956): 20–63.

Neugebauer, O. *Mathematical Cuneiform Texts.* New Haven: American Oriental Society, 1945.

Neugebauer, O. *The Exact Sciences in Antiquity.* Princeton: Princeton University Press, 1952.

———. *Vorlesungen über Geshichte der antiken mathematischen Wissenschaften con O. Neugebauer.* Berlin, New York: Springer-Verlag, 1969.

Novikov, S. P. "Topology in the 20th century: A View from the Inside." *Russian Mathematical Surveys* 59 (2004): 3–28.

O'Connor, J. J., and E. F. Robertson. The MacTutor History of Mathematics Archive, http://www-history.mcs.st-andrews.ac.uk/history/index.html.

O'Shaughnessy, P. *A Case of Lies.* New York: Random House, 2005.

Osserman, R. *Poetry of the Universe.* New York: Doubleday, 1995.

Otal, J.-P. "Thurston's hyperbolization of Haken manifolds," in *Surveys in Differential Geometry,* edited by C. C. Hsiung and S.-T. Yau. Cambridge: International Press, 1998, vol. 3, 77–194.

Papakyriapoulos, C. D. "On Dehn's Lemma and the Asphericity of Knots." *Proceedings of the National Academy of Sciences* 43 (1957): 169–72 and *Annals of Mathematics* 66 (1957): 1–26.

———. "On Solid Tori." *Proceedings of the London Mathematical Society* 3, Ser. 7 (1957): 281–99.

———. "Some Problems on 3-Dimensional Manifolds." *Bulletin of the American Mathematical Society* 64 (1958): 317–35.

———. "A Reduction of the Poincaré Conjecture to Group Theoretic Conjectures." *Annals of Mathematics* 77 (1963): 250–305.

Parshall, K. H., and D. E. Rowe. *The Emergence of the American Mathematical Community 1876–1900: J. J. Sylvester, Felix Klein, and E. H. Moore.* Providence: American Mathematical Society, 1994.

Parshall, K. H., and A. C. Rice, eds. *Mathematics Unbound: The Evolution of an International Mathematical Research Communtiy, 1800–1945.* Providence: American Mathematical Society, and London: London Mathematical Society, 2002.

Perelman, G. "Proof of the Soul Conjecture of Cheeger and Gromoll." *Journal of Differential Geometry* 40 (1994): 299–305.

———. "The Entropy Formula for the Ricci Flow and Its Geometric Applications." math.DG/0211159 (11 November 2002), "Ricci Flow with Surgery on Three-Manifolds." math.DG/0303109 (10 March 2003), "Finite Extinction Time for the Solutions to the Ricci Flow on Certain Three-Manifolds." math.DG/0307245 (17 July 2003).

Peterson, M. "Dante and the 3-Sphere," *American Journal of Physics* 47 (1979): 1031–1035.

Poincaré, H. "Sur les Functions Fuchsiennes." *Comptes rendus de l'Académie des sciences* 92 (February 14, 1881): 333–35; 92 (February 21, 1881): 395–96; 92 (April 4, 1881): 859–61.

———. *Oeuvres de Henri Poincaré.* Paris: Gauthiers-Villars, 1952.

———. *Papers on Fuchsian Functions.* Translated by J. Stillwell. New York: Springer-Verlag, 1985.

———. *New Methods of Celestial Mechanics.* Edited and introduced by D. Goroff. New York: American Institute of Physics, 1993.

———. *The Value of Science: Essential Writings of Henri Poincaré.* Edited by Stephen Jay Gould. New York: The Modern Library (Random House), 2001.

Pont, J.-C. *La Topologie Algébrique des origins à Poincaré.* Paris: Presses Universitaires de France, 1974.

Ptolemy, C. *The Geography.* Edited and translated by E. L. Stevenson. New York: Dover, 1991.

Rêgo, E., and C. Rourke. "Heegaard Diagrams and Homotopy 3-Spheres." *Topology* 27 (1988): 137–43.

Reid, C. *Hilbert.* New York, Heidelberg, Berlin: Springer-Verlag, 1970.

———. *Courant in Göttingen and New York, The Story of an Improbable Mathematician.* New York, Heidelberg, Berlin: Springer-Verlag, 1976.

Riemann, B. *Gesammelte mathematische Werke und wissenschaftlicher Nachlass.* (Zweite Auflage, bearbeitet von Heinrich Weber), Leipzig: Teubner, 1892, 541–58.

Rourke, C. "Characterisation of the Three-Sphere following Haken." *Turkish Journal of Mathematics* 18 (1994): 60–69.

———. "Algorithms to Disprove the Poincaré Conjecture." *Turkish Journal of Mathematics* 21 (1997): 99–110.

Rowe, D. E. "'Jewish Mathematics' at Göttingen in the Era of Felix Klein." *Isis* 77 (1986): 422–49.

Rowe, D. E. "Klein, Hilbert and the Göttingen Mathematical Tradition." *Osiris* (1989): 186–213.

Russell, J. B. *Inventing the Flat Earth: Columbus and Modern Historians.* Westport: Praeger Publishing, 1991.

Russo, L. *The Forgotten Revolution: How Science Was Born in 300 BC and Why It Had to Be Reborn,* New York: Springer, 2003.

Saari, D. G., ed. *The Way It Was: Mathematics from the Early Years of the* Bulletin, Providence: American Mathematical Society, 2003.

Sabbagh, K. *The Riemann Hypothesis: The Greatest Unsolved Problem in Mathematics.* New York: Farrar, Straus, and Giroux, 2002.

Sarkaria, K. S. "The Topological Work of Henri Poincaré." In *History of Topology,* edited by I. M. James. Amsterdam: Elsevier, 1999, 123–68.

Scholz, E. "The Concept of a Manifold, 1850–1950," in *History of Topology,* edited by I. M. James. Amsterdam: Elsevier, 1999, 25–64.

Seifert, H., and W. Threlfall. *A Textbook of Topology.* Translated by M. A. Goldman. New York: Academic Press, 1980.

Smale, S. "The Generalized Poincaré Conjecture in Dimensions Greater than Four." *Annals of Mathematics* 74 (1961): 391–406.

Sossinsky, A. *Knots: Mathematics with a Twist.* Translated by G. Weiss. Cambridge, MA: Harvard University Press, 2002.

Stallings, J. "On the Loop Theorem." *Annals of Mathematics* 72 (1960): 12–19.

Stallings, J. R. "How Not to Prove the Poincaré Conjecture," in *Topology Seminar Wisconsin,* 1965. Edited by R. H. Bing. Princeton: Princeton University Press, 2006, 83–88.

Stillwell, J. *Geometry of Surfaces.* New York: Springer-Verlag, 1992.

Thickstun, T. L. "Taming and the Poincaré Conjecture." *Transactions of the American Mathematical Society* 238 (1978): 385–96.

———. "Open Acyclic 3-Manifolds, a Loop Theorem and the Poincaré Conjecture." *Bulletin of the American Mathematical Society* 4 (1981): 192–94.

Thom, R. *Mathematical Models of Morphogenesis.* Translated by W. M. Brookes. New York: Halsted Press, 1983.

Threlfall, W., and H. Seifert. "Topologische Untersuchung der Discontuitätsbereiche endlicher Bewegungsgruppen der dreidimensionalen sphärische Raumes I." *Mathematische Annalen* 104 (1930): 1–70.

Thurston, W. P. "Existence of Codimension-one Foliations." *Annals of Mathematics* 104 (1976): 249–68.

———. "Mathematical Education." *Notices of the American Mathematical Society* 37 (1990): 844–50.

———. "On Proof and Progress in Mathematics." *Bulletin of the American Mathematical Society* 30 (1994): 161–77.

———. *Three-Dimensional Geometry and Topology*, vol. 1. Edited by S. Levy. Princeton: Princeton University Press, 1997.

———. "How to See 3-Manifolds," *Classical and Quantum Gravity* 15 (1994): 2545–71.

———. *The Geometry and Topology of Three-Manifolds*. Electronic version 1.1, March 2002, http://www.msri.org/publications/books/gt3m.

Tietze, H. "Über die topologischen Invarianten mehrdimensional Mannigfaltigkeiten." *Monatshefte für Mathematik und Physik* 19 (1908): 1–118.

Veblen, O. *Analysis Situs*. New York: American Mathematical Society, 1922.

van der Waerden, B. L. *Science Awakening I: Egyptian, Babyloniann, and Greek Mathematics*. Translated by A. Dresden with additions by the author. Leyden: Noordhooff, 1975.

Weber, C., and H. Seifert. "Die beiden Dodekaedräume." *Mathematische Zeitschrift* 37, no. 2 (1933) p. 237.

Weeks, J. *The Shape of Space*, 2nd ed. New York, Basel: Marcel Dekker, 2002.

———. "The Poincaré Dodecahedral Space and the Mystery of the Missing Fluctuations." *Notices of the American Mathematical Society* 51 (2004) 610–619.

Weil, A. "Riemann, Betti, and the Birth of Topology." *Archive for History of Exact Sciences*, 20 (1979): 9–96.

Whitehead, A. N. *The Aims of Education and Other Essays*. New York: Macmillan, 1967.

Whitehead, J. H. C. "Certain Theorems about Three-Dimensional Manifolds." *Quarterly Journal of Mathematics* 5 (1934): 308–20.

———. "Three-Dimensional Manifolds (Corrigendum)." *Quarterly Journal of Mathematics* 6 (1935): 80.

———. "On the Sphere in 3-Manifolds." *Bulletin of the American Mathematical Society* 64 (1958): 161–66.

ARCHIVES

Archives Henri Poincaré, L'Université de Nancy 2. (Much of this very useful archive is online: http://www.univ-nancy2.fr/poincare/.)

Centre Historique des Archives Nationales, Paris. Cote AJ/16/6124.

Archives, Academie des Sciences Paris.

Universitätsarchiv Göttingen (http://wwwuser.gwdg.de/~uniarch/).

Further Reading

Here is a very short list of suggestions for further reading. For full citations, consult the bibliography.

For biographies of mathematicians, the best reference is Gillispie's *Dictionary of Scientific Biography*. The online MacTutor History maintained by J. J. O'Connor and E. F. Robertson (www-history.mcs.st-andrews.ac.uk/history/index.html) contains lively, short biographies of many mathematicians. It also contains well-written, accessible summaries of a number of mathematical topics. Wikipedia (http://en.wikipedia.org/wiki) continues to get stronger and more scholarly.

For individual biographies, see Laugwitz on Riemann, Buhler on Gauss, and Reid on Hilbert. Parts of Laugwitz and Bühler require strong mathematical backgrounds. At the time of writing, there is no critical biography of Poincaré. One is promised in 2010. In the meantime, Bottazzini's summary of Poincaré's life and work is very readable and highly recommended. It is only available in French, however. Galison's book on the social context that influenced Poincaré and Einstein's work on time is superb. For more on the social organizations that support science and mathematics, see Nakayama's book.

For readers with minimal mathematical backgrounds, the best reference on the relationship between geometry and topology in the case of two- and three-dimensional manifolds is Jeff Weeks's *The Shape of Space*. This book does not require calculus.

For those with more background, Thurston's book, *Three-Dimensional Geometry and Topology*, and his Princeton lecture notes (available on MSRI's Web site) are wonderful.

The essay by Cannon, et al. in Levy's book *Flavors of Geometry* has an introduction to a number of different models of hyperbolic geometry that is easily accessible if you have had a calculus course.

For very readable introductions to knot theory, see the books of C. Adams or A. Sossinsky, preferably both (as they cover different material).

For readers with a couple of years of college calculus, a good short introduction to differentiable topology is Kosinski's book *Differentiable Manifolds*. Barden and Thomas also do a nice job. For Riemannian geometry, I recommend Gallot, Hulin, and Lafontaine. These books require a serious commitment on the part of the reader (and, preferably, someone else to work through them with).

Hamilton's paper 1995 survey article on singularities of the Ricci flow (reprinted in Cao, et al.'s *Collected Papers on the Ricci Flow*) is probably the best place to start for more information on the Ricci flow. The forthcoming book by Tian and Morgan will have a full, refereed exposition of Perelman's work and the proof of the Poincaré conjecture.

Art credits

Figures 1 and 2 modified from the CIA's *The World Factbook* (www.cia.gov/cia/publications/factbook).

Figure 3: Library of Congress (B. Agnese, *Portolan atlas of 9 charts and a world map, etc. Dedicated to Hieronymus Ruffault, Abbot of St. Vaast,* 1544, Call No. G1001.A4 1544, digital ID:g3200m gct00001, http://hdl.loc.gov/loc.gmd/g3200m.gct00001).

Figures 4 and 5 modified from *The World Factbook.*

Figure 20: Gustav Doré's sketch of the Empyrean from Dante's *Divine Comedy.*

Figure 27: Oil painting of Carl Friedrich Gauss by C. A. Jensen (1792–1870). Original in the Archiv der Georg-August-Universität, Göttingen, Germany.

Figures 39 (Riemann), 40 (Klein), and 49 (Hilbert): Archiv der Stiftung Benedictus Gotthelf Teubner Leipzig/Dresden/Berlin/Stuttgart, http://www.stiftung-teubner-leipzig.de/. Note that Riemann was dead in 1868, so the date cannot refer to the date the photograph was taken.

Figures 26 and 41: Photography by Mary O'Shea.

Figure 42 (Poincaré): Smithsonian digital collection, http://www.sil.si.edu/digitalcollections/hst/scientific-identity/explore.htm.

Figure 50 (Hamilton): International Congress of Mathematicians, http://www.icm2006.org.

Figure 53 (Möbius band): Science Photo Library.

All other diagrams created by the author, modified by, and used by permission of Rizzoli.

The lyrics of the song "Lobachevsky" on pages 72–73 are used with permission of their author, Tom Lehrer.

Acknowledgments

The book grew out of a conversation over a bottle of wine some summers ago in the backyard of the home of my brother, Stephen, and his wife, Jill Pearlman. Steve, who was up to his eyeballs in medieval Christian-Muslim battles, asked what was new in mathematics. I told him of Grisha Perelman's Internet posting and lectures, their curious reception, the Poincaré conjecture, and the possibility that the arguments would hold up. Steve insisted that the story would make a good book and that I could write it. He said he at least would read it, and a publisher, George Gibson, might be interested. Steve read successive drafts, and made many, many suggestions even as he struggled to launch his latest book, *Sea of Faith*. I cannot thank him enough.

I owe an even larger debt to my wonderful Mary, wife, best friend, and fun fellow traveler. It was she who suggested taking a sabbatical to start the book. She was its first reader and a ruthless critic, systematically excising pretension and self-indulgent verbiage. She took the pictures in figure 41, after having talked to the pharmacist on the ground floor of what was briefly Poincaré's childhood home. He graciously allowed us to enter the courtyard and roam through the upstairs. Mary was the model for figure 35. My children, too, were wonderfully supportive critics. Seamus and Sarah read the manuscript at a very early stage and made suggestions that led to a complete reframing of the opening chapters. Brendan and Kathleen read parts, often asked how it was going, and lent enthusiastic support. I thank, too, my brother, Kevin, my father, Mary's father, and members of my extended family for support and, in some cases, critical feedback.

I have been blessed with extraordinary colleagues at Mount Holyoke College and in the mathematical community. Joanne Creighton, president of the college and English professor, encouraged me, made it possible for me to take a sabbatical leave, and read the manuscript. Penny Gill, professor of Politics, served as acting dean of faculty during my leave. She and Sally Sutherland, my associate dean and a Shakespeare scholar, worked long hours covering the hours I spent away from my desk. Sally and my assistant Susan Martin also supplied wonderful logistical support. David Cox of Amherst College, Harriet Pollatsek and Nicole Vaget of Mount Holyoke, and Ron Davidoff read an embarrassingly early draft of the manuscript and vastly improved it. Andy Lass and George Cobb provided detailed comments on a less embarrassing, but far from final, draft. Thanks to Lester Senechal, Mark Peterson, Jane Crosthwaite, Giuliana Davidoff, Jillian McLeod, and Char Morrow for sympathetic readings. I am also grateful to James Carlson, Alan Durfee, Peter Lax and Jeff Weeks, whose comments saved me from many errors. Needless to say, the remaining errors are mine alone.

I would probably never have even considered undertaking this project had it not been for an invitation from Chris Benfey and Karen Remmler, then codirectors of Mount Holyoke's Weissman Center for Leadership and the Liberal Arts, to speak and write for nonmathematicians about the French mathematician Jacques Hadamard, for a symposium. Although writing about mathematics for a lay audience was harder than I thought, I found that I enjoyed it. I also discovered that I really enjoyed archival work. Karen and Chris encouraged my nascent interest and provided useful comments about both that earlier paper and a draft of this book. Bob Schwartz provided invaluable help in navigating French archives. A faculty grant from the college underwrote some of the archival research for this project. Such grants, and the support from the endowed chair that I hold, would not be possible without the support and commitment of the college's alumnae.

I am very grateful to the mathematics departments at the University of Miami and the University of Edinburgh for their warm hospitality in the fall 2004 and spring 2005 semesters, respectively. Miami's department chair, Alan Zame, and his assistants, Dania Puerto and Toni Taylor, were very helpful. At Edinburgh, Elmer Rees provided help, support, mathematical guidance, and encouragement.

I have had a great deal of technical help. Archivists Édith Pirio in the Centre historique des Archives nationales (Paris), Florence Greffe at the Academie des Sciences (Paris), and Ulrich Hunger at the Niedersächsische Staats- und Universitätsbibliothek Göttingen (Göttingen) were incredibly helpful. Some of the material I had requested on Poincaré was not easily available because it was filed with records of another person. The prospects of making another trip back to Paris to obtain the records nearly defeated me. Without prompting, Édith Pirio took it upon herself to photocopy them, and mailed them to me in Edinburgh. I cannot thank her adequately. Eli Gottlieb provided both detailed line-by-line editing and macro-editing of an early rough draft. Ed Hernstadt advised me on contractual matters. George Gibson and Jackie Johnson of Walker & Company have been terrific. Jackie's editing, in particular, has been truly extraordinary and to the point.

I thank my parents, Anne and Daniel, who worked hard to afford my brothers, Kevin and Stephen, and me educations that they themselves did not have. They created a loving home in which nothing was impossible and where ideas flowed freely. The first two chapters of this book were written in Ottawa's Montfort hospital while sitting with my father at my mother's deathbed. I think she would have liked the result, and wish she were alive to read it.

Index

A Note on the Author

Donal O'Shea is dean of faculty and vice president for Academic Affairs at Mount Holyoke College, where he also is the Elizabeth T. Kennan Professor of Mathematics. He has written scholarly books and monographs, and his research articles have appeared in numerous journals and collections. O'Shea is a member of the American Mathematical Society, the Mathematics Association of America, and mathematical societies of Canada, London, and France. He lives in South Hadley, Massachusetts.